Julia Bohnert

Effects of Time-Varying Magnetic Fields in the Frequency Range 1 kHz to 100 kHz upon the Human Body

Numerical Studies and Stimulation Experiment

Vol. 15
Karlsruhe Transactions on Biomedical Engineering

Editor:
Karlsruhe Institute of Technology
Institute of Biomedical Engineering

Eine Übersicht über alle bisher in dieser Schriftenreihe erschienenen Bände finden Sie am Ende des Buchs.

Effects of Time-Varying Magnetic Fields in the Frequency Range 1 kHz to 100 kHz upon the Human Body

Numerical Studies and Stimulation Experiment

by
Julia Bohnert

KIT Scientific Publishing

Dissertation, Karlsruher Institut für Technologie
Fakultät für Elektrotechnik und Informationstechnik, 2011

Impressum

Karlsruher Institut für Technologie (KIT)
KIT Scientific Publishing
Straße am Forum 2
D-76131 Karlsruhe
www.ksp.kit.edu

KIT – Universität des Landes Baden-Württemberg und nationales
Forschungszentrum in der Helmholtz-Gemeinschaft

KIT Scientific Publishing 2012
Print on Demand

ISSN: 1864-5933
ISBN: 978-3-86644-782-0

Effects of Time-Varying Magnetic Fields in the Frequency Range 1 kHz to 100 kHz upon the Human Body - Numerical Studies and Stimulation Experiment

Zur Erlangung des akademischen Grades eines

DOKTOR-INGENIEURS

von der Fakultät für

Elektrotechnik und Informationstechnik

des Karlsruher Instituts für Technologie (KIT)

genehmigte

DISSERTATION

von

Dipl.-Ing. Julia Bohnert

geb. in Oberkirch (Baden)

Tag der mündlichen Prüfung:	10. November 2011
Referent:	Prof. Dr. rer. nat. Olaf Dössel
Korreferent:	Prof. Dr. rer. nat. Thorsten M. Buzug

Danksagung

Diese Arbeit entstand während meiner Tätigkeit als wissenschaftliche Mitarbeiterin am Institut für Biomedizinische Technik des Karlsruher Instituts für Technologie. Einige Personen haben mich in dieser Zeit auf verschiedene Weise unterstützt, denen ich im Folgenden danken möchte.

Zuallererst bedanke ich mich bei Herrn Prof. Dr. rer. nat. Olaf Dössel für die Gelegenheit in seiner Arbeitsgruppe zu arbeiten, für die zahlreichen Anregungen und die immer wieder interessanten wissenschaftlichen Diskussionen, die maßgeblich zum Gelingen dieser Arbeit betgetragen haben.

Herrn Prof. Dr. rer. nat. Thorsten M. Buzug möchte ich für sein Interesse an der Arbeit und die Übernahme des Korreferates ebenfalls herzlich danken.

Für ihre hilfreichen Tipps und Anregungen, sowie ihre konstruktive Kritik bei Konferenzbeiträgen und Veröffentlichungen bedanke ich mich bei den Mitarbeiten vom Philips Forschungslabor in Hamburg; im Besonderen Dr. Jörn Borgert, Bernhard Gleich, Dr. Ingo Schmale und Dr. Jürgen Rahmer.

Ein besonderer Dank gilt der Deutsche Telekom Stiftung, von der ich nicht nur durch die finanzielle Unterstützung sehr profitiert habe. Das Angebot an Seminaren und Workshops, sowie die Kontakte, die sich ergeben haben, habe ich in meiner Stipendiatenzeit sehr geschätzt. Die Teilnahme am 61. Lindauer Nobelpreisträgertreffen, die mir die Deutsche Telekom Stiftung ermöglicht hat, bleibt ein unvergessliches Erlebnis.

Dem gesamten Team des Studienzentrums für Sehgeschädigte, allen voran Joachim Klaus, danke ich, dass es mir auch über das Studium hinaus mit beratender und tatkräftiger Unterstützung zur Seite stand.

Des Weiteren möchte ich den Studenten danken, die mit ihrer Arbeiten einen Beitrag zum Gelingen dieses Projekts geleistet haben. Ohne sie wäre manches Projekt im Sand verlaufen oder manche Idee nie geboren worden.

Auch meinen Kollegen und allen weiteren Mitarbeitern des Instituts für Biomedizinische Technik möchte ich danken für die vielen interessanten und oft auch sehr unterhaltsamen Gespräche, sowie ihre Unterstützung, wo sie notwendig war.

Natürlich geht ein großer Dank an meine Freunde und vor allem an meine Familie, die trotz schwieriger Zeiten Verständnis und Interesse für meine Arbeit gezeigt hat. Danke an Jasmin, die immer das Richtige sagt und immer ein Trostpflaster parat hat. Und nicht zuletzt danke ich meinem Freund Christian für alles, was er ist.

Karlsruhe, im November 2011 Julia Bohnert

Contents

Part I Background

Part II Investigating Field Effects upon Excitable Tissue

Part I

Background

1 Introduction

1.1 Motivation

In 2005, PHILIPS RESEARCH in Hamburg, Sector Medical Imaging Systems, introduced a new imaging technique, Magnetic Particle Imaging (MPI). MPI promises fast acquisition of high resolution images in both time and space of internal body functions. Therefore, exhibited magnetic nano-particles are excited by oscillating magnetic fields. The magnetization of these particles provides the imaging information.

The effect of time-varying magnetic fields on the human body between 1 kHz and 100 kHz are yet poorly investigated. The new imaging technique Magnetic Particle Imaging uses magnetic fields of about 25 kHz. The development of a new medical imaging system is complex and costly. Hence the developers need to know about unwanted side-effects before the first human is exposed to the generated fields. From electromagnetic field theory, it is known that time-varying magnetic fields induce rotating electric fields. Depending on the tissue's conductivity, an electric current is induced that might be able to stimulate peripheral nerves and muscles, or — even worse — the heart. Stimulating the heart, in part or as a whole, may lead to life-threatening events of fibrillation. Stimulation of peripheral muscles may be tolerable considering the profit of high resolution images promised by Magnetic Particle Imaging, if certain thresholds are not exceeded.

It is also known, that electric stimulation requires higher field amplitudes when the field frequency increases. The ability of excitation finally stops, but the "cut-off"-frequency is not known yet. Furthermore, the field power is turned into heating power. Hence, the energy deposit might lead to local or global body warming. In order to assure safety for the patients undergoing MPI diagnosis, the biological effects of the applied magnetic fields are so far ambiguous and need to be investigated.

1.2 State of the Art

Thanks to the development of Magnetic Resonance Imaging (MRI), algorithms for tissue segmentation and classification, the number of computer models for simulations of healthy and pathologic scenarios is growing. Powerful computer hard- and software allow complex simulations of field distributions.
Furthermore, with the advancement of MRI, communications technologies, cancer treatment by localized heat generation and magneto-stimulation therapies, the need for understanding the effects of electromagnetic fields on biological tissue is continuously growing. Studies of electromagnetic compatibility of high frequency electromagnetic fields have been carried out extensively during the last decades. Numerical simulations present a powerful means for investigation of field distributions in numerous applications.
Several numerical studies have been carried out regarding the effects of MRI gradient coil switching [1–4]. The reliable results of numerical simulation also give hope, that people with implants may undergo MRI diagnosis [5].
In 1998, the group of P. J. DIMBYLOW presented induced current densities from low-frequency magnetic fields in a 2 mm resolution model of the body [6]. He applied the scalar potential finite difference (SPFD) method for field calculations up to 10 MHz. This is so far the only publication found that deals with numerical studies of low frequency magnetic fields affecting the body as a whole.

1.3 Aim of this Work

This work was a subproject of the joint research project *MAGIC*, that had originally been conducted by the German Federal Ministry of Education and Research (BMBF), initiated by PHILIPS RESEARCH. In this work, the field distributions resulting from exposure to MPI field generating coils were to be quantified and evaluated with respect to stimulating and heating effects in humans. In order to estimate the behavior of excitable tissue over frequency, the frequency range up to 100 kHz was covered in the studies. A human body model was employed for numerical calculations using the Finite Element Method. The behavior of cellular structures was also investigated by means of numerical simulations. Finally, simulated results were compared with results of a magnetic stimulation experiment.
During the time this thesis was established, the development of Magnetic Particle Imaging continuously progressed. The field parameters that will be used in this project are the state of research of 2010.

1.4 Structure of the Thesis

This thesis is made up of three parts: The first part comprises the first and second chapters which introduce the reader to the topic and present the most important background knowledge underlying to this project. These cover the principle concepts of Magnetic Particle Imaging, including some details about scanner hardware and image reconstruction. A brief look into electromagnetic field theory is given in Section 2.2, where the basic physics of time-varying magnetic fields are dealt with. The Finite Element Method, which is used for field calculations in this thesis, is briefly explained there, too. The following section explains electric mechanisms of the cell membrane of excitable tissue with an introduction to electrophysiological modeling. Section 2.4 deals with the known effects of magnetic fields on biological tissue and how they can be modeled mathematically. The chapter closes with considerations and previous work upon stimulation thresholds and derived guidelines for limiting exposure to time-varying fields.

Part II comprises three chapters that describe the motivation, methods and results of the three core topics that have been processed in this thesis. The first one, handled in Chapter 3, deals mainly with electro-magnetic and thermal simulations on a human torso model exposed to a set of field generating coils. Furthermore, the measurement and simulation of temperature increase in an MPI laboratory scanner is described that was carried out for evaluation of human body thermal simulation results. The effect of induced current densities on the transmembrane voltages of excitable tissues is investigated in Chapter 4. Finally, a stimulation experiment is presented in Chapter 5. The aim of the experiment was the determination of stimulation thresholds for magnetic stimulation by means of a small figure-of-eight coil. The complete process of hardware configuration, experiment procedure and results are explained and visualized.

A summary, final discussion and conclusions are given in Part 3. An approach has been made to relate the different projects to each other and to evaluate consistency with former findings.

2 Fundamentals

In this chapter, some essentials will be provided about underlying technical and physiological backgrounds. It starts with a short introduction to Magnetic Particle Imaging, its basic principle and image acquisition theory. Second, a brief look into electromagnetic field theory is given, which is focused on the kHz range which this whole work is about. This section contains one approach of Finite Element Method, which presents a common method for numerical solution of partial differential equations. The following section deals with cellular electrophysiology of excitable tissues and how the cell membrane's functionality can be modeled mathematically. After having described some field theory and basic electrophysiological processes of cell excitation, effects of time-varying magnetic fields on biological tissue are explained with focus on magnetic stimulation and thermal heating. Finally, a brief overview is given of findings upon stimulation behavior of excitable cells which influenced the design of guidelines for limiting exposure to time-varying electric and magnetic fields.

2.1 Magnetic Particle Imaging

Over the last decades, tomographic imaging methods such as computer tomography (CT) [7, 8], magnetic resonance imaging (MRI) [9]) and positron-emission tomography (PET) [10] have become essential tools for the diagnosis of a large number of diseases. PET provides high sensitivity in static imaging based on tracer materials, while MRI and CT provide images of excellent tissue contrast at high spatial resolution. MRI, CT and PET are also capable of dynamic imaging. Using MRI, magnetic tracers can be followed in the patient's body offering the chance of functional and bio-molecular imaging. However, despite recent advances, real-time imaging of 3D volumes using MR and magnetic tracers remains challenging. For instance, the main restricting factor in the use of magnetic tracers in MRI is

the detection threshold. MRI image acquisition is based on resonant excitation of proton spin magnetization. Transforming measured data into a spatial image requires a trade-off between good signal to noise ratio and good spatial or temporal resolution. CT offers higher temporal resolution, but applies ionizing radiation. Magnetic Particle Imaging (MPI) promises fast 3D images of magnetic tracers both high spatial and temporal resolution.
The possible medical applications of MPI are manifold: The most probable in terms of feasibility and tracer development success are cardio-vascular applications: MPI could be utilized to determine cardiac ejection fraction and flow dynamics, quantitative myocardial perfusion and vulnerable plaque detection. Other medical fields could be interventional radiology, neuro-vascular applications, organ perfusion imaging, gastro-intestinal applications, oncology, breast imaging, and cell tracking.
The following description of MPI is based on publications of the PHILIPS research group in Hamburg, released during the years 2005 to 2009 [11–14]. A summary of a variety of MPI related research, including hardware realization and tracer development, is given in [15].

2.1.1 Principle Concept

MPI uses the nonlinearity of the magnetization curves of superparamagnetic materials and their saturation behavior at some magnetic field strength. If an oscillating magnetic field $H(t)$ is applied to magnetic particles, the magnetic material will exhibit a magnetization $M(t)$ according to the material's specific magnetization curve $M(H)$. If the magnetic field does not exceed a specific saturation level, the particles will exhibit vast changes in magnetization on the linear part of their magnetization curve (see Fig. 2.1(a)). If, in addition, a *static* magnetic field of high amplitude is applied to the magnetic particles, they saturate and the oscillating magnetic field does not effect in further particle magnetization (Fig. 2.1(b)).
In MPI, a strong time-constant field gradient is applied which saturates magnetic particles in most of the observed region, leaving a field free point (FFP) in the very center. This field is called *selection field.* By superimposing a time-varying magnetic field, particles within the FFP will produce a signal containing the driving frequency f and harmonics of f, in terms of induced voltages in some receive coils. These higher frequencies are separated from the received signal by means of appropriate filtering. Outside the FFP, harmonics are suppressed which allows easy spatial encoding of the received signal. By steering the FFP through the vol-

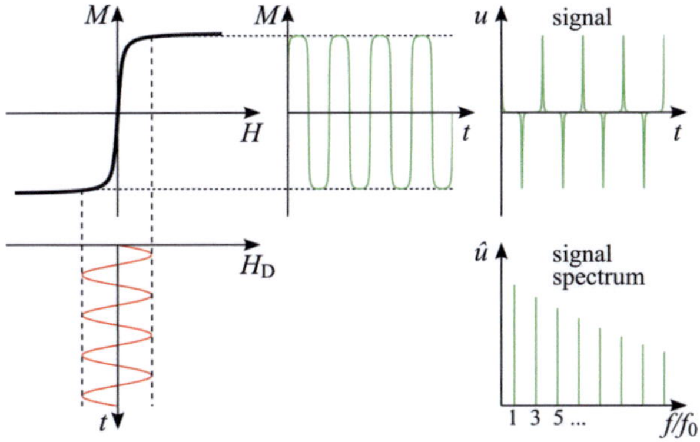

(a) Excitation of particles at the position of the field free point

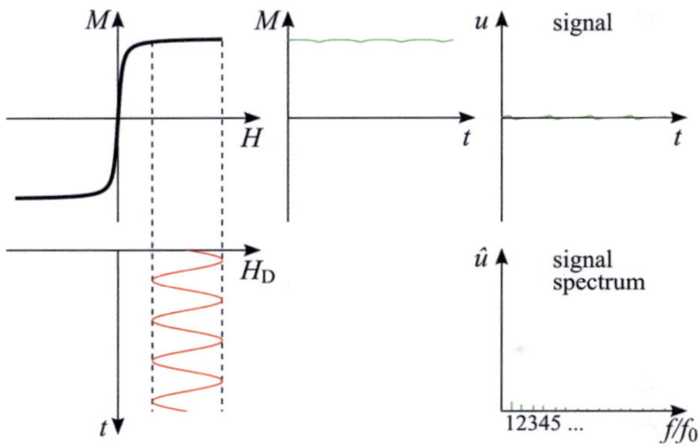

(b) Excitation of particles outside the position of the field free point

Fig. 2.1. Response of magnetic particles to an external time-varying magnetic field. **(a):** When applying an oscillating magnetic field (H), the magnetic material exhibits a magnetization $M(t)$. Since $M(H)$ is non-linear, the recorded signal u includes higher harmonics, as is shown in the signal's spectrum $\hat{u}$. **(b):** In case of an additional time-constant field, the oscillating field does not significantly change the magnetization of the material, as it is already in saturation. Almost no harmonics are evident in the signal.

ume of interest, a tomographic image can be generated [11]. Moving the FFP is achieved by three orthogonal homogenous magnetic fields, called *drive fields*. Thus, the method is capable of generating real-time high quality images of the spatial distribution of magnetic nanoparticles.
This description is best suited to explain the principle of MPI. In a further step it can be recognized that the sinusoidal sweep through the magnetization curve is actually not needed. Moving the FFP very fast through the volume of interest will already be sufficient for imaging purpose.

2.1.2 First 3D Scanner Hardware

After the introduction of the new image modality in 2005 [11], first in-vivo 3D real-time MPI scans were presented in 2009 [14]. The method was developed well enough to reveal details of a beating mouse heart using a clinically approved concentration of a commercially available MRI contrast agent. The temporal resolution achieved was 21.5 ms at a 3D field of view of 20.4 mm × 12 mm × 16.8 mm with a spatial resolution sufficient to resolve all heart chambers.

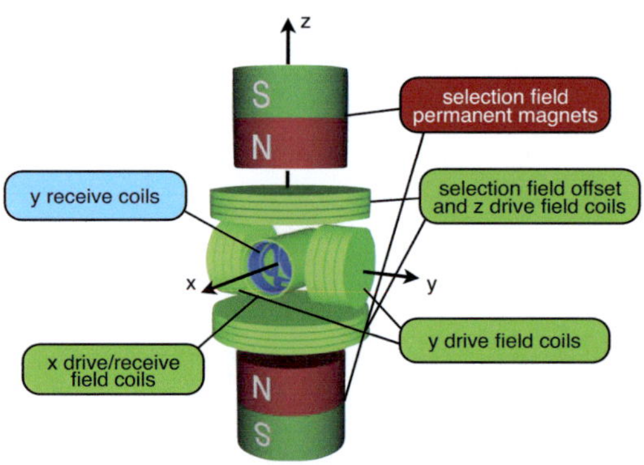

Fig. 2.2. Schematic scanner setup. Two permanent magnets and the coil pair in z direction generate the selection field. The drive field coils move the FFP in all three spatial directions. Each spatial component of the magnetization is detected by a respective receive coil. In x direction, the drive field coil is also used for signal reception. The bore diameter is 32 mm.

In this first 3D MPI scanner (Fig. 2.2), the selection field was realized by two permanent magnets, whose north poles were facing each other, plus a pair of electric magnets, hereby achieving a field gradient of 5.5 T/m in the center. The higher the

magnetic field gradient, the smaller the FFP. It must be considered though, that the drive field needs to compensate the selection field at the edges of the field of view (FOV) in order to achieve an FFP there. Additionally, for a human size scanner it might be costly to realize such high fields.
The drive field was generated by three coil pairs, moving the FFP on a Lissajous trajectory. For 3D imaging, the drive field frequencies were set to 2.5 MHz/102, 2.5 MHz/99 and 2.5 MHz/96, covering a field of view with a center amplitude of 18 mT. The Lissajous trajectory had a repetition time of 21.5 ms, corresponding to encoding 46.42 volumes per second. The size of the gaps in the Lissajous pattern was chosen to match the desired resolution in the order of 1 mm [14]. The Lissajous trajectory was chosen for the sake of simplicity; various possible trajectories have been assessed since then [16]. The faster the FFP moves over the object, the higher the local rate of change in the particles' magnetization and hence the higher the induced signal recorded by the receive coils. Two saddle-type receive coil pairs were aligned approximately perpendicularly to the bore. In the axial direction, the solenoid drive field coil was also used for receiving the signal. The resulting bore hole was 32 mm.

2.1.3 Image Acquisition

For image reconstruction, the functional $||\mathbf{GC}-\mathbf{U}||^2+\lambda||\mathbf{C}||^2$ has to be minimized. Here, the matrix **G** represents the system function[1], which was acquired beforehand, **U** is the measured data, **C** is the desired image, and λ is a regularization parameter chosen to adjust the balance between signal-to-noise ratio (SNR) and resolution for best visual image impression [14]. To this end, a row relaxation method called *algebraic reconstruction technique (ART)* [18] was used [14].
From the received data, only harmonics with amplitudes above the noise floor of the system function are used in the reconstruction, and all these amplitudes are normalized to the same level. The very same normalization factors are applied to the object data, too [12]. Thus, the signal's bandwidth used for reconstruction goes from $\approx$ 50 kHz to $\approx$ 1 MHz. This upper cut-off frequency is where the signal dropped below noise level [13].

2.1.4 Tracer Characteristics

The choice of the employed magnetic particles contributes much to the achieved image resolution. Bigger particles show a steeper magnetization curve which in-

[1] Usually, the term "system function" is used for linear, time-invariant systems. However, the system function of MPI depends on both location and frequency [17].

creases spatial resolution. This effect is limited between 40 nm and 50 nm, where no significant improvement is detectable [13].
The magnetization curves of magnetic nanoparticles follow the LANGEVIN-function (see also Fig. 2.1):

$$\mathbf{M}(\mathbf{r},t) = M_0(\mathbf{r}) \left[\coth\left(\frac{m|\mathbf{B}(\mathbf{r},t)|}{k_B T}\right) - \frac{k_B T}{m|\mathbf{B}(\mathbf{r},t)|}\right] \mathbf{e}_\mathbf{B}(\mathbf{r},t), \tag{2.1}$$

with

$$M_0(\mathbf{r}) = \frac{N(\mathbf{r})m}{\Delta V}. \tag{2.2}$$

The quantities are:

$\mathbf{B}(\mathbf{r},t)$	:	total magnetic field representing the sum of the selection field and the drive fields, at position $\mathbf{r}$ and time t
m	:	magnetic moment of the nanoparticle
k_B	:	BOLTZMANN constant
T	:	absolute temperature
$\mathbf{e}_\mathbf{B}(\mathbf{r},t)$	:	the unit vector in the field direction
$M_0(\mathbf{r})$	:	saturation magnetization of the particles contained in the volume ΔV
ΔV	:	size of the volume elements
N	:	number of particles each contributing a magnetic moment m

From the LANGEVIN function it is derived that particles are characterized by their magnetic moment $m = \frac{1}{6}\pi D^3 M_s$ and thus by their diameter D and saturation magnetization M_s.
Particle concentration also contributes to the achieved resolution, since more particles deliver more signal. The lower the concentration, the stronger becomes the noise background in relation to the signal [13]. In MRI, a usual tracer dosage is 8 μmol(Fe)/kg whereas the approved safe dosage is 40 μmol(Fe)/kg for human applications.
In the first experiments of 2009, the applied tracer for MPI was *Resovist*, a conventional MRI contrast agent from SCHERING, Berlin. The applied concentrations ranged between 8 μmol(Fe)/kg, and 45 μmol(Fe)/kg, which is slightly above the safe dosage for human applications [14]. Since there is a trend towards new MRI tracers, *Resovist* is not available anymore. Those new materials consist of Gadolinium complexes which present not enough signal harmonics due to their magnetic behavior. Hence, much effort is put into tracer development for MPI, with the aim at larger particle diameters and lower size distributions. Figure 2.3 shows one

frame out of a video in which the tracer bolus can be followed passing through all heart chambers of a narcotized mouse. To relate the MPI signal to the mouse anatomy, reference MR images of the mouse were acquired after the MPI scans.

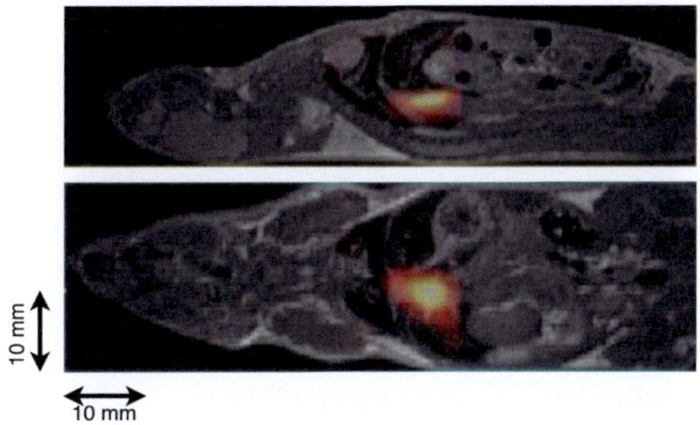

Fig. 2.3. Sample of in-vivo MPI scan [14]. The FOV was placed to fit the complete mouse heart. Static MRI data were fused with dynamic MRI images. Here, the tracer bolus just enters the FOV via the vena cava. One volume is acquired in 21.5 ms enabling to video the tracer passage through all heart chambers without triggering or motion decompensation [14].

The spatial resolution in the scan of the beating mouse heart was estimated to be 1 mm in the z-direction and 2 mm in the orthogonal directions. At the edges of the FOV, the resolution decreased to 1.5 mm and 3 mm, respectively. Image resolution is not homogeneous in the total FOV. This results from differing selection field gradients depending on the gradient direction. The decrease in resolution towards the edges of the FOV results from slower FFP movement and hence a smaller induced signal.

Simulations regarding system function, feasible resolution and SNR of a potential human MPI scanner have already been carried out [13]. The expected SNR was estimated to be 52 % of the SNR found in the mouse system.

Much effort is put into new tracer development, since a lot of improvement in SNR, resolution, speed and image quality lies in the tracer's characteristics, namely in the steepness of the magnetization curve and harmonics generation.

However, it has been shown that the MPI modality already provides images of sufficient resolution for many medical applications. Still, besides tracer development, there is much room for improvement e. g. in encoding sequences and reconstruction algorithms [12–14].

To cover a larger volume than the current FOV, an additional field similar to the drive field, but at a lower frequency ($\leq$ 100 Hz), can be applied. This "focus field" can move the volume covered by the drive fields to any volume of interest.

2.2 A Brief Look into Electromagnetic Theory

The following sections give a very brief insight into electromagnetic field theory. Contents of these sections are extracted from well known literature. It is not intended to give a lecture here, but for the sake of completeness, basic laws and ideas shall be reproduced. Emphasis is put on issues that are relevant for this thesis. Formulas and methods are extracted from [19–22]. Literature of electromagnetic theory is manyfold.

2.2.1 The Nature of Electric and Magnetic Fields

JAMES CLERK MAXWELL put together the laws of electromagnetism in the form of four equations. These formulas describe physical laws all over the frequency spectrum. Table 2.1 summarizes the equations in integral form and in differential form.

Differential form	Integral form	Remarks
$\nabla \cdot \mathbf{D} = \rho$	$\oint \mathbf{D} \cdot d\mathbf{a} = \int \rho dv$	GAUSS's law
$\nabla \cdot \mathbf{B} = 0$	$\oint \mathbf{B} \cdot d\mathbf{a} = 0$	GAUSS's law of magnetic fields (Non-existence of isolated magnetic charge)
$\nabla \times \mathbf{E} = -\frac{\partial \mathbf{B}}{\partial t}$	$\oint \mathbf{E} \cdot d\mathbf{l} = -\frac{\partial}{\partial t} \oint \mathbf{B} \cdot d\mathbf{a}$	FARADAYS's law
$\nabla \times \mathbf{H} = \mathbf{J} + \frac{\partial \mathbf{D}}{\partial t}$	$\oint \mathbf{H} \cdot d\mathbf{l} = \oint \left(\frac{\partial \mathbf{D}}{\partial t}\right) \cdot d\mathbf{a}$	AMPERE's circital law

Table 2.1. Generalized forms of MAXWELL's equations. **E**, **D**, **H**, **B** and **J** represent the electric field, displacement current, magnetic field, magnetic flux density and the free current density field, respectively. ρ denotes the distribution of free charges.

Other physical laws, like the LORENTZ force equation ($\mathbf{F} = q(\mathbf{E} + \mathbf{v} \times \mathbf{B})$) or the equation of continuity ($\nabla \cdot \mathbf{J} = -\frac{\partial \rho_V}{\partial t}$) go hand in hand with MAXWELL's equations.

The characteristic behavior of electric and magnetic fields, however, differs significantly for different frequencies. To be more exact, the behavior of fields depends on the relation between the size L of the observed domain and the wavelength λ of the field. If $\lambda \gg L$, quasi-static electromagnetic theory applies. The electric and magnetic fields are *uncoupled*. The $\frac{\partial}{\partial t}$-terms in MAXWELL's third and fourth equations (FARADAY's law and AMPERE's law) are negligible. High frequency characteristics apply if $\lambda \ll L$. This is the frequency range where the term *electromagnetic fields* is appropriate. **E** and **B** are related to each other such that the temporal change of one is the source of the other.

For convenience, the term "EM-fields" used here refers to the complete fields all over the frequency range, although in the low frequency band, the term "electric fields and magnetic fields" is more appropriate.

Important characteristics of EM-fields are how they interact with materials and how the materials interact with the fields. **E** and **B** are defined to account for forces among charges. When referring to field-material-interactions the topic is about forces on charges. Interactions with materials are described by three effects:

- induction of dipole polarization
- alignment of already existing electric dipoles
- movement of free charges

Induced polarization and alignment of permanent dipoles is accounted for by the material's permittivity ε. Movement of free charges refers to conduction currents and is accounted for by the material's ohmic conductivity σ. All three effects are accounted for in AMPERE's law:

$$\nabla \times \mathbf{H} = \mathbf{J_c} + j\omega\tilde{\varepsilon}\tilde{\mathbf{E}} = \sigma_c\tilde{\mathbf{E}} + j\omega(\varepsilon_{real} - j\varepsilon_{imag})\tilde{\mathbf{E}} \tag{2.3}$$

Movement of charges, free or bounded, is lossy, whereas dipole oscillation is partly lossless. Usually, the term "conductivity" actually refers to what is here denoted as $\sigma_c + \omega\varepsilon_{imag}$. The part $j\omega\varepsilon_{real}\mathbf{E}$ is the displacement current, which represents the lossless portion of the oscillation of the bound charges. Hence a complex permittivity is defined as

$$\tilde{\varepsilon} = \varepsilon' - j\varepsilon'' = \varepsilon_{real} - j\left(\frac{\sigma_c}{\omega} + \varepsilon_{imag}\right) \tag{2.4}$$

Finally, the permeability μ describes how much partial alignment of permanent magnetic dipoles occurs for a given applied **B**. Similar to the complex permittivity, a complex permeability $\tilde{\mu}$ is defined, where the real part of $\tilde{\mu}$ describes the lossless interaction between the magnetic field and the magnetic dipoles of the material,

while the imaginary part represents any loss associated with aligning the magnetic dipoles.
With ε, μ and σ, the following constitutive relationships between EM-fields and material properties are defined:

$$\mathbf{D} = \varepsilon_0 \mathbf{E} + \mathbf{P}$$
$$\mathbf{B} = \mu_0 (\mathbf{H} + \mathbf{M})$$
$$\mathbf{J} = \sigma \mathbf{E}$$

P and **M** summarize electric and magnetic polarization effects, respectively. In case of linear and isotropic materials, these formulations boil down to $\varepsilon \mathbf{E}$ and $\mu \mathbf{H}$. In order to satisfy MAXWELL's equations, the following relationships can be derived for the behavior of EM-fields at a boundary between two materials 1 and 2:

$$E_{1t} = E_{2t} \tag{2.5}$$
$$H_{1t} - H_{2t} = K \tag{2.6}$$
$$D_{1n} - D_{2n} = \rho_s \tag{2.7}$$
$$B_{1n} = B_{2n} \tag{2.8}$$

K is the surface current and ρ_s surface charges. Indices t and n denote tangential and normal parts of respective field vectors.
Another way of expressing electrodynamics is completely possible even without electric and magnetic fields. These lead to the potential formulation including the electric potential (or scalar potential) ϕ and the magnetic potential (or vector potential) **A** defined as follows:

$$\mathbf{E} = -\nabla \phi - \frac{\partial \mathbf{A}}{\partial t} \tag{2.9}$$
$$\mathbf{B} = \nabla \times \mathbf{A} \tag{2.10}$$

With these definitions, FARADAY's Law and GAUSS's law for magnetism are automatically satisfied and the other two give the following equations:

$$\nabla^2 \phi + \frac{\partial}{\partial t}(\nabla \cdot \mathbf{A}) = -\frac{\rho}{\varepsilon_0} \tag{2.11}$$

$$\left(\nabla^2 \mathbf{A} - \frac{1}{c^2}\frac{\partial^2 \mathbf{A}}{\partial t^2}\right) - \nabla\left(\nabla \cdot \mathbf{A} + \frac{1}{c^2}\frac{\partial \phi}{\partial t}\right) = -\mu_0 \mathbf{J} \tag{2.12}$$

In electro(quasi)statics, Eq. (2.11) yields the POISSON equation:

$$\Delta\phi = f \tag{2.13}$$

with Δ being the LAPLACE operator $\nabla\cdot\nabla$ or ∇^2 and f being any function describing a charge distribution divided by ε_0.
Assuming harmonic oscillations $e^{j\omega t}$, it is usually more convenient to transform MAXWELL's equations into frequency domain. Now, **E**, **D**, **H**, **B** and **J** are the phasors of the electric field, displacement current, magnetic field, magnetic flux density and the current density field, respectively. In the frequency domain, $\partial/\partial t \equiv j\omega$. As described above, the complex permittivity $\tilde{\varepsilon}$ is defined as $\tilde{\varepsilon} := \varepsilon_0\varepsilon_r + \frac{\sigma}{j\omega}$. The frequency domain MAXWELL equations read:

$$\nabla\times\mathbf{E} = -j\omega\mathbf{B} \tag{2.14}$$

$$\nabla\times\mathbf{H} = \mathbf{J} + j\omega\mathbf{D} \tag{2.15}$$

$$\nabla\cdot\mathbf{D} = \rho \tag{2.16}$$

$$\nabla\cdot\mathbf{B} = 0 \tag{2.17}$$

The charge continuity equation follows directly from Equations (2.15) and (2.16):

$$\nabla\cdot\mathbf{J} + j\omega\rho = 0 \tag{2.18}$$

2.2.2 Low Frequency Approximations

In case the wavelength λ is long compared to the computational domain, electric fields and magnetic fields are decoupled. Mathematically, this is expressed by means of the vector potential **A** with $\nabla\times\mathbf{A} = \mathbf{B}$. This solves Equations (2.14) and (2.17) *(*HELMHOLTZ *decomposition)*:

$$\mathbf{E} = -j\omega\mathbf{A} - \nabla\phi = \mathbf{E_s} + \mathbf{E_i} \tag{2.19}$$

where E_s is *s*olenoidal (= divergence-free) and E_i is *i*rrotational (= curl-fee). Substitution of **B** in Eq. (2.15) written by the vector potential yields

$$\nabla\times\frac{1}{\mu}\nabla\times\mathbf{A} = \omega^2\tilde{\varepsilon}\mathbf{A} - j\omega\tilde{\varepsilon}\nabla\phi + j_0 \tag{2.20}$$

and the charge continuity equation becomes

$$\nabla\cdot\tilde{\varepsilon}\nabla\phi = j\omega\nabla\cdot\tilde{\varepsilon}\mathbf{A}. \tag{2.21}$$

This is valid when no charges are generated, ($\nabla\cdot j_0 = 0$), e.g., in case of closed current loops.
Eq. (2.20) reduces to

$$\nabla \times \frac{1}{\mu} \nabla \times \mathbf{A} = j_0 \tag{2.22}$$

in case all materials satisfy the quasi-static condition[2]:

$$d \ll \lambda \quad \Leftrightarrow \quad |\omega^2 \tilde{\varepsilon} \mu d^2| \ll 1 \tag{2.23}$$

Now, the electric field and the magnetic field are completely decoupled with the vector potential $\mathbf{A}$ being the magneto static vector potential $\mathbf{A_0}$.
Furthermore, if μ is constant over the entire computational domain Ω, $\mathbf{A_0}$ is calculated by the law of BIOT-SAVART:

$$\mathbf{A_0}(\mathbf{r}) = \frac{\mu}{4\pi} \int_{\Omega} \frac{\mathbf{j_0}(\mathbf{r}')}{|\mathbf{r} - \mathbf{r}'|} d^3\mathbf{r}' \tag{2.24}$$

Depending on the given problem, the respective equations are solved:

- In case $\mathbf{j_0} = 0$, the vector potential $\mathbf{A}$ in Eq. (2.21) vanishes and the electric field only consists of the scalar potential ϕ *(electro quasi-static)*:

$$\nabla\cdot \tilde{\varepsilon} \nabla \phi = 0 \tag{2.25}$$

- In case $\mathbf{j_0} \neq 0$, Eq. (2.22) states that the vector potential is the magneto static vector potential $\mathbf{A_0}$ "excited" by the current distribution $\mathbf{j_0}$. Therefore, $\mathbf{A_0}$ is calculated by solving the magneto static problem (2.22) or, if $\mu = const$, in Ω, by integrating Biot-Savart (Eq. (2.24)). Then, the **E** field is determined by using again Eq. (2.21) and the fact that $\nabla \cdot \mathbf{j_0} = 0$ (closed current loops):

$$\nabla \cdot \tilde{\varepsilon} \nabla \phi = j\omega \nabla \cdot (\tilde{\varepsilon} \mathbf{A_0}) \qquad \textit{(magneto quasi-static)} \tag{2.26}$$

Both complex-valued problems, electro quasi-static (2.25) and magneto quasi-static (2.26), can be further simplified to real valued problems,

[2] In general, $\mu = \mu_0 \mu_r$. Here it is supposed to be constant all over the computational domain ($\mu_r = 1$)

	electro quasi-static ($\mathbf{j_0} = 0$)	magneto quasi-static ($\mathbf{j_0} \neq 0$)
$\sigma \ll \omega\varepsilon \Leftrightarrow \tilde{\varepsilon} \rightarrow \varepsilon$	electro static	displacement currents dominated
$\sigma \gg \omega\varepsilon \Leftrightarrow \tilde{\varepsilon} \rightarrow \sigma$	stationary currents	ohmic currents dominated

if one condition holds for all materials within the computational domain.

2.2.3 A Technique for Numerical Calculations of Fields: The Finite Element Method

In general, electric, magnetic and electromagnetic fields in biological tissue cannot be assessed analytically without major simplifications of the given problem. Computer assisted numerical methods are powerful but often of high complexity. The most common methods used depending on the specific problem are:

- Finite Element Methods
- Finite Difference Methods
- Charge Simulation Methods
- Boundary Element Methods
- Method of Moments
- Monte Carlo Methods

Most of these methods convert the underlying functional equation into equations of matrices. In many technical applications one faces boundary value problems which can be described by the LAPLACE equation ($\Delta\phi = 0$). The difficulty lies in putting down the boundary conditions. Sometimes the boundary conditions are accounted for by converting the boundary value problem into a POISSON problem. The homogeneous equation of this field problem then is the LAPLACE equation with the trivial solution 0 [20].
In *SEMCAD X*, low frequency problems are solved using the Finite Element Method (FEM). FEM is a suitable method for solving problems that can be described by partial differential equations. This section will glance at the subsequent steps during the calculation process. Solving a problem by means of the FEM, basically takes five steps:

1. discretization of the computational domain into a finite number of subregions or *elements*.
2. approximation of the potential distribution within one element
3. derivation of governing equations or element matrices
4. assembly of all elements in the solution region

5. consideration of the boundary conditions and solution of equations obtained.

Explaining FEM is easiest by using an example. Therefore, a two-dimensional region problem is considered in which the potential distribution is sought for.
The POISSON equation is an elliptic partial differential equation (PDE) of second order:

$$-\nabla^2\phi = \frac{\rho}{\varepsilon} \text{ in } \Omega \tag{2.27}$$

In the Variational Method, this partial differential equation can be translated into the problem of minimizing the functional:

$$W_\Omega = \int_\Omega W_e(x,y)d\Omega = \int_\Omega \frac{1}{2}\varepsilon\mathbf{E}^2(x,y)d\Omega \tag{2.28}$$

With $\mathbf{E} = -\nabla\phi(x,y) = -(\frac{d\phi}{dx} + \frac{d\phi}{dy})$, Eq. (2.28) becomes

$$W_\Omega = \frac{1}{2}\varepsilon \int_\Omega \left(\left(\frac{\partial\phi}{\partial x}\right)^2 + \left(\frac{\partial\phi}{\partial y}\right)^2 \right) d\Omega. \tag{2.29}$$

In physics, an equilibrium is always achieved when the stored energy is at its minimum. This would require $\frac{\partial W}{\partial \phi} \stackrel{!}{=} 0$

Finite Element Discretization

In the first step, the region Ω is divided into a number of — usually triangularly-shaped (other shapes are possible as well) — non-overlapping elements (Figure 2.4). In practice, it is important that elements are more or less similar and that acute angels are avoided.

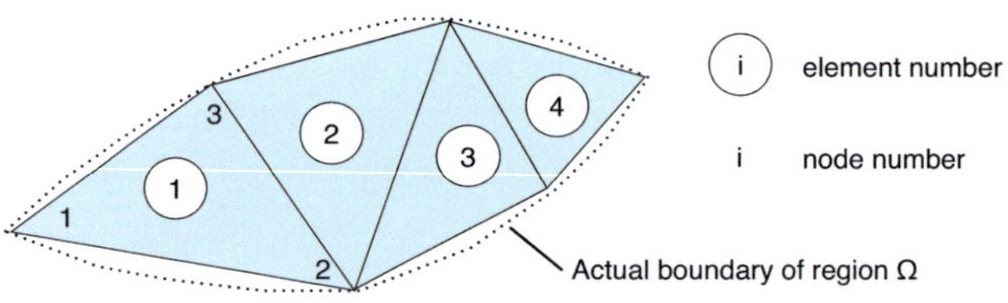

Fig. 2.4. A typical finite element subdivision of an irregular domain

Within each element e an approximation for the (local) potential ϕ_e is sought. Thereafter, the potential distribution will be inter-related such that the potential is continuous across inter-element boundaries.

Approximation Function $\phi_{\Delta e}(x,y)$ within one Element

The approximation solution for the whole region is:

$$\phi(x,y) \simeq \sum_{e=1}^{N} \phi_e(x,y), \tag{2.30}$$

where N is the number of triangular elements.
The simplest way to create an approximation function for a 1D region are linear piecewise defined functions, or *hat functions*. In case of triangular elements, the function for ϕ within an element e is a polynomial approximation.

$$\phi_{\Delta A}(x,y) = a + bx + cy \begin{cases} \neq 0, & \text{if } x,y \in A, \\ = 0, & \text{otherwise.} \end{cases} \tag{2.31}$$

In that case elements do only interact with direct neighbors with whom they share two nodes.

Element Governing Equations

Now, Eq. (2.31) shall be solved within element e. Therefore, the potential ϕ is determined at the nodes e_1, e_2 and e_3 of the triangle.

$$\begin{bmatrix} \phi_{e_1} \\ \phi_{e_2} \\ \phi_{e_3} \end{bmatrix} = \begin{bmatrix} 1\ x_1\ y_1 \\ 1\ x_2\ y_2 \\ 1\ x_3\ y_3 \end{bmatrix} \cdot \begin{bmatrix} a \\ b \\ c \end{bmatrix} \tag{2.32}$$

Solving this for the coefficients a, b, and c yields a function of node potentials and coordinates of the nodes:

$$a,b,c = f(\phi_1,\phi_2,\phi_3,x_1,x_2,x_3,y_1,y_2,y_3) \tag{2.33}$$

Substitution of the coefficients into the original approximation function (2.31) yields a new version of the approximation function, in which the potential $\phi_{\Delta A}(x,y)$ is obtained out of three nodal potentials by interpolation.

$$\phi_{\Delta A}(x,y) = \sum_{n=1}^{3} \alpha_n(x,y)\phi_n \tag{2.34}$$

The functions α_n are called *interpolation functions* or *element shape functions* and depend on the shape and location of the element. Using a linear approximation, α_n are calculated by:

$$\alpha_i(x,y) = \frac{1}{2\Delta A}\left((x_j y_k - x_k y_j) + (y_j - y_k)x + (x_k - k_j)y\right) \tag{2.35}$$

with circularly changing indices for α_j and α_k, respectively. ΔA, the area of element e is given as follows:

$$2\Delta A = \begin{vmatrix} 1 & x_1 & y_1 \\ 1 & x_2 & y_2 \\ 1 & x_3 & y_3 \end{vmatrix} \tag{2.36}$$

The value of ΔA is positive if the nodes are numbered counterclockwise.

Derivation of Governing Equations or Element Matrices

Various techniques exist for determination of the element matrices. Here, the *variational formulation* will be outlined. Other techniques for example are the *method of weighted residuals* (Galerkin) or the *energy balance method.*
Applying the ∇-operator on Eq. (2.34) yields

$$\nabla\phi_{\Delta A}(x,y) = \sum_{n=1}^{3} \nabla\alpha_n\phi_n. \tag{2.37}$$

Going back to the electric energy per unit length s associated with element e (while $\rho_s = 0$) and substituting (2.37) leads to:

$$w_e = \frac{1}{2}\int \varepsilon|\mathbf{E}|^2 d\mathbf{s} \tag{2.38}$$

$$= \frac{1}{2}\int \varepsilon|\nabla\phi_e|^2 d\mathbf{s} \tag{2.39}$$

$$= \frac{1}{2}\sum_{i=1}^{3}\sum_{j=1}^{3} \varepsilon\phi_{ei}\left[\int \nabla\alpha_i \cdot \nabla\alpha_j d\mathbf{s}\right]\phi_{ei} \tag{2.40}$$

By replacing the term in brackets by $C_{ij}^{(e)}$, Eq. (2.40) can be written in matrix form as

$$w_e = \frac{1}{2}\varepsilon{\phi_\mathbf{e}}^T\mathbf{C}^{(\mathbf{e})}\phi_\mathbf{e}. \tag{2.41}$$

The matrix

$$\mathbf{C^{(e)}} = \begin{bmatrix} C_{11}^{(e)} & C_{12}^{(e)} & C_{13}^{(e)} \\ C_{21}^{(e)} & C_{22}^{(e)} & C_{23}^{(e)} \\ C_{31}^{(e)} & C_{32}^{(e)} & C_{33}^{(e)} \end{bmatrix}$$

is usually called the *element coefficient matrix* (or *stiffness matrix* in structural analysis). Its elements C_{ij} reflect the coupling between nodes i and j. Consequently $\mathbf{C^{(e)}}$ is symmetric.
One element in $\mathbf{C}$ is for example:

$$C_{12}^{(e)} = \int \nabla\alpha_1 \cdot \nabla\alpha_2 \, d\mathbf{s} \tag{2.42}$$

$$= \frac{1}{4A^2}[(y_2 - y_3)(y_3 - y_1) + (x_3 - x_2)(x_1 - x_3)] \int d\mathbf{s} \tag{2.43}$$

$$= \frac{1}{4A}[(y_2 - y_3)(y_3 - y_1) + (x_3 - x_2)(x_1 - x_3)] \tag{2.44}$$

System Matrix Assembly

Finally, after consideration of one element e within the region Ω, all elements need to be assembled. The energies of all elements can be summed up to a total of

$$W = \sum_{e=1}^{N} w_e = \frac{1}{2}\varepsilon\phi^T\mathbf{C}\phi \tag{2.45}$$

with $\phi^T = [\phi_1 \dots \phi_n]$, n the number of nodes and N the number of elements. $\mathbf{C}$ is called the *global coefficient matrix* whereas $\mathbf{C^{(e)}}$ was the element coefficient matrix. The elements C_{ij} are obtained by using the fact that the potential distribution must be continuous across interelement boundaries. The contribution to the i,j position in $\mathbf{C}$ comes from all elements containing nodes i and j. $\mathbf{C}$ is symmetric and singular. $C_{ij} = 0$ if nodes i and j are decoupled. Hence, $\mathbf{C}$ becomes sparse and banded. This makes it comparably easy to solve.

Solving the System

As stated earlier, the system is satisfied in the state of minimal energy stored within the solution region Ω is minimum. Hence, all partial derivatives of W with respect to each node value of the potential must be 0:

$$\frac{\partial W}{\partial \phi_k} = 0 \quad k = 1, 2, \dots, n \tag{2.46}$$

This leads to

$$0 = \sum_{i=1}^{n} \phi_i C_{ik}. \tag{2.47}$$

Considering all nodes, this can be put into matrix form:

$$\mathbf{C}\phi = 0. \tag{2.48}$$

Hence, a set of simultaneous equations is obtained from which the potentials ϕ_n can be found. One way of solving the system is the *band matrix method*. The goal is to bring the system into the form

$$\mathbf{Ax} = \mathbf{b}, \tag{2.49}$$

since there are many mathematical methods for solving such kind of problems. This is the point where the boundary values come into play: Within the region of interest Ω, free nodes, i.e. nodes of unknown potential, are numbered first and fixed nodes, meaning nodes of known potential, are numbered last. Since the potentials at fixed points are constant, Eq. (2.48) can be rewritten to

$$\mathbf{C_{ff}}\phi_\mathbf{f} = -\mathbf{C_{fp}}\phi_\mathbf{p} \tag{2.50}$$

with indices f and p referring to nodes with unknown potentials and known (or prescribed) potentials, respectively. Solving (2.50) for $\phi_\mathbf{f}$ yields the desired form of $\mathbf{Ax} = \mathbf{B}$ which can be solved either directly (in case of small problems) or iteratively. Solving this problem is not part of the FEM.

2.3 An Excerpt of Cell Physiology

2.3.1 Physiology of Excitable Tissue

The ability of specialized cells to answer a stimulation by a change in membrane potential or generation on action potential is called excitability. Thus, excitable tissue refers to nerve, muscle, and certain secretor cells. The latter are not subject in this thesis.

The body's information system is based on electric impulses conducted from sender to receiver. A sender can be any location on nerves and the receiver can be other nerves, muscle tissue or neurons in the brain. Responsible for the ability of excitation is the distribution of ions in the cell environment.

2.3.1.1 Nernst Equation and Resting Potential

The cell membrane separates the intracellular medium to the extracellular medium. Both media consist mainly of water (80 %). Intra- and extracellular space contain various ions of different concentrations, which evoke a transmembrane voltage (TMV). Primal cations are K^+-, Na^+- and Ca^{2+}-ions. Primal anions are Cl^-- and $HCO_3{}^-$-ions. Like nutrients and metabolic substances, ions pass the cell membrane via specific tunnels and pumps. The cell membrane's characteristic is responsible for the ion distribution. The resting potential, the status in which the cell is inactive, is the result of balancing the electric force due to lack of protons inside the cell and chemical force due to the ambition towards concentration equilibrium. Both forces are opposed to each other. This equilibrium potential V of each ion type X is expressed by the NERNST-equation, in consideration of the intra- and extracellular concentration ($[X]_i$ and $[X]_e$) and the ion's charge number q [23]:

$$V_X = \frac{RT}{Fq} \ln \frac{[X_e]}{[X_i]} \tag{2.51}$$

with R being the gas constant (8.314 J/(mol·K)), F the FARADAY constant ($9.65 \cdot 10^4$ As/mol) and T the absolute temperature in K.
The state of force equilibrium results in the resting potential, that is described in the GOLDMANN-HODGKIN-KATZ equation [24]:

$$V_m = \frac{RT}{Fq} \ln \frac{P_{Na} \cdot [Na^+]_e + P_K \cdot [K^+]_e + P_{Cl} \cdot [Cl^-]_i}{P_{Na} \cdot [Na^+]_i + P_K \cdot [K^+]_i + P_{Cl} \cdot [Cl^-]_e} \tag{2.52}$$

with P being the membrane permeability for the respective ion. Table 2.2 summarizes human ion concentrations and respective NERNST potentials.

ion	intracellular $[X_i]$ [mmol/l]	extracellular $[X_e]$ [mmol/l]	relation	permeability P	NERNST potential [mV]
Na^+	7 – 11	144	1:12	0.03	60 – 80
K^+	120–155	4 – 5	30:1	0.07	-91 – -95
Cl^-	4 – 7	120	1:20	0.9	-74 – 82
Ca^{2+}	$10^{-5} - 10^{-4}$	1.3	$> 10^4 : 1$		126.5 – 157.3

Table 2.2. Typical ion concentrations and resulting NERNST potentials in human skeletal muscle cells [23, 25]

2.3.1.2 Action Potential Generation and Signal Propagation

If the ion concentrations are shifted, due to an externally applied field for example, ion channels react to the change accordingly. If the extracellular space gets more positive, the cell membrane is hyperpolarized, in contrast to getting depolarized, if the outside gets more negative. As soon as the depolarization exceeds the threshold potential, an action potential is triggered. An action potential is the result of ion channels opening and closing which leads to characteristic inflow and outflow of ions.

At the beginning, the Na^+ permeability of the cell membrane increases very rapidly, allowing sodium ions to flow from outside to inside, which finally leads to a positive transmembrane voltage. The Na^+ channels close again immediately. After that, the more slowly increasing potassium ion permeability allows K^+ ions to flow from inside to outside, thus returning the intracellular potential to its resting value. The maximum excursion of the membrane voltage during activation can amount up to 100 mV. In cardiac cell, the release of Ca^{2+} elongates the action potential, which prevents high frequent repetition of action potential generation. After the resting potential is reestablished, the Na-K pump restores the ion concentrations inside and outside the membrane to their original values.

In neural cells, electrical signals propagate along the nerve fiber until they reach the destined receiver which might be another neural cell, a muscle cell, or a neuronal cell in the brain. The signal to the receiver cell is transferred electrically or biochemically. In case of a motoneuron, the adjacent muscle cell well decode the received signal into mechanical contraction, whose strength and duration depends on the number and frequency of incoming signals. Skeletal muscle cells do not send signals to neighboring muscles, in contrast to cardiac muscle cells. Cardiac cells are interconnected via *gap junctions*; membrane proteins which allow fast excitation propagation between cardiac muscle cells. Excitation of few cardiac cells usually causes the contraction of the atria and/or ventricles, depending on the location of the stimulus.

Shape, duration and involved ions at the cell membrane of excitable tissue differ depending on the cell type. Figure 2.5 shows characteristic action potentials of a neural and a cardiac cell. Neural action potentials last about 1 ms, whereas the cardiac action potential duration is up to 400 ms. The elongation is caused by Ca^{2+} inflow into cardiac cells. The plateau phase prevents fast repetitions of cell depolarization which would lead to reentry effects and disturb the natural excitation propagation. The resting potential of cardia cells is ≈ -85 mV in contrast to -70 mV of neural cells.

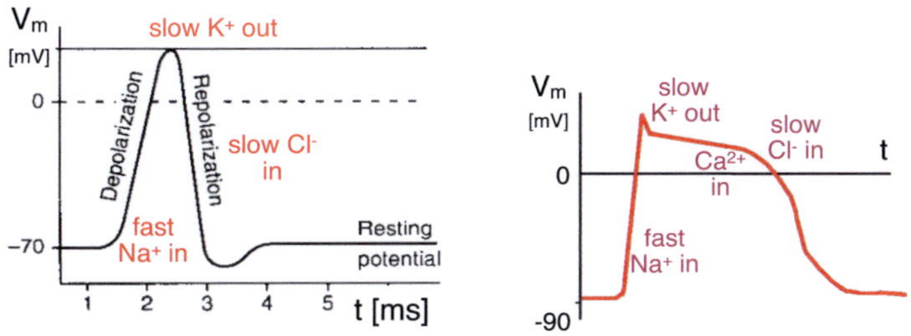

(a) Membrane voltage (V_m) during a propagating nerve impulse

(b) Action potential of a cardiac cell: The plateau is generated by an intracellular Ca^{2+} release

Fig. 2.5. Action potential shapes and ion flows of neural and cardiac cells

2.3.1.3 Dielectric Properties of Biological Tissue

Dielectric properties of biological tissues are frequency-dependent and *dispersive*. Their typical behavior is characterized by three distinctly large dielectric dispersions, usually referred to as α-, β-, and γ-dispersions [26]. The α-dispersion occurs below 10 Hz. Some tissues express a drop of their relative permittivity by some decades. The frequency range from tens of kHz to tens of MHz is the range of β-dispersion. There, the capacitive characteristic of the cell membrane becomes evident. With increasing frequency, the reactance of the membrane decreases which leads to an increase of effective conductivity. In addition, the ε_r falls due to less charging capability. The γ-dispersion is dominant in the microwave frequency range. The cell membrane is completely short-circuited. Permittivity continuously decreases which is caused by the polar characteristics of water-molecules.
Dielectric properties have been formulated mathematically by expressing the exponential course of molecule depolarization. Depending on the field frequency $f = \omega/2\pi$ and the relaxation constant τ, the complex relative permittivity $\tilde{\varepsilon}$ can be written as

$$\varepsilon'_r(\omega) = \varepsilon'_{r\infty} + \sum_n \frac{\varepsilon'_{rs,n} - \varepsilon'_{r\infty}}{1 + (j\omega\tau_n)^{1-\alpha_n}} + \frac{\sigma_s}{j\omega\varepsilon_0^2}, \tag{2.53}$$

implying that $\varepsilon'_{r\infty}$ is the permittivity of the tissue at infinite frequency, and ε'_{rs} the permittivity of the tissue of very low frequency. Every term n of the summation (DEBYE-term) accounts for one dispersion range. σ_s represents the static ion conductivity and α an empiric parameter introduced by COLE AND COLE [27]. If the constants $\varepsilon'_{rs,n}$, $\varepsilon'_{r\infty}$, τ_n, α_n and σ_s of a tissue are given, its dielectric properties can be determined with Eq. (2.53). In 1996, GABRIEL ET AL. summarized those constants of 50 different biological tissues [28]. The effective electric conductivity σ and the relative permittivity ε_r can now be determined by the COLE-COLE equation with constants of GABRIEL ET AL. and decomposition in real part and imaginary part according to the definition of the complex permittivity (Eq. (2.4)). Figure 2.6 shows the results for dielectric properties σ and ε_r of skeletal muscle tissue, determined by Eq. (2.53)[3]. However, GABRIEL values are subject to a great variation and reflect only an average or estimate for those tissues of which no measured date are available.

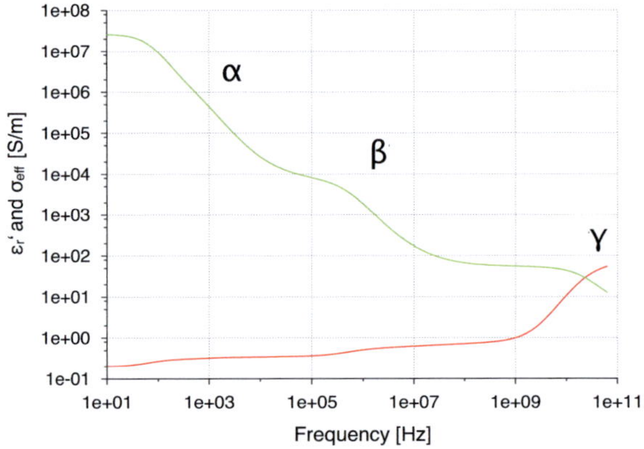

Fig. 2.6. Electric conductivity (red curve) and relative permittivity (green curve) of muscle tissue [29]

GABRIEL values refer to properties at 37°C. However, dielectric properties are also temperature dependent. σ and ε_r rise by ≈ 2 %/°C and 1.5 %/°C for frequencies up to 500 MHz and 1 GHz, respectively, and decrease there above. Values for temperature dependency below 100 kHz are rare and insecure. Hence, dielectric

[3] In this thesis, the term GABRIEL values refers to the results of the COLE-COLE equation for effective conductivity and relative permittivity.

properties are mostly treated as temperature independent in thermodynamic calculations below 100 kHz [29].

2.3.2 Electrophysiological Modeling

In order to understand the mechanisms of action potential generation and propagation, mathematical models of excitable cells, especially myocardial cells have long been developed. In 1970, BEELER AND REUTER published a model of the electrical activity of the mammalian ventricular myocyte [30]. The model was based on the formation introduced by HODGKIN AND HUXLEY in 1952, who had described transmembrane ionic currents of the squid's nerve cell upon experimental data [31]. With the development of single-cell and single-channel recording techniques, the intracellular and extracellular ionic environments, channel kinetics and membrane ionic currents could be assessed and incorporated into mathematical description.
All these models are built by a series of coupled differential equations, which describe the non-linear, resistive and capacitive behavior of the cell. Regarding the cell membrane as a capacitor connected in parallel with variable resistances and batteries representing the different ionic currents and pumps, the rate of change of the membrane potential can be written as [30–35]:

$$\frac{dV}{dt} = -\frac{1}{C_m}(I_{ion} + I_{st}) \tag{2.54}$$

where C is the membrane capacitance, I_{ion} the sum of ionic currents and I_{st} the externally applied stimulus current. In general, the current flow I of ion x can be derived from

$$I_x = g_x(V - V_x) \tag{2.55}$$

while V is the actual potential and V_x results from the equilibrium potential of the respective ion. g_x represents the membrane conductance for the ion x. g_x changes depending on the state of the respective ion gate. Single ion gates can either be *open* or *closed*. The transition is described with *rate constants* α and β, which are voltage dependent. If n represents the (relative) number of open gates and $1-n$ the (relative) number of closed gates, the change of the number of open gates can be written as

$$\frac{dn}{dt} = \alpha_n(1-n) - \beta_n n \tag{2.56}$$

If a steady state is reached, eq. (2.56) becomes 0, which leads to the number of open gates in the steady state n_∞ to be:

$$n_\infty = \alpha_n/(\alpha_n + \beta_n) \tag{2.57}$$

With that, eq. (2.56) can also be written as

$$\frac{dn}{dt} = \frac{n_\infty - n}{\tau_n} \tag{2.58}$$

or, replacing n by y, calling y the *gating variable*:

$$\frac{dy}{dt} = \frac{y_\infty - y}{\tau_y} \tag{2.59}$$

where τ_y represents the time constant which describes the temporal behavior of a gate returning to the steady state after some kind of disturbance:

$$\tau_y = \frac{1}{\alpha_y + \beta_y} \tag{2.60}$$

With that, for example the sodium current I_{Na} is formulated as

$$I_{Na} = G_{Na_{max}} m^3 hj(V - V_{Na}) \tag{2.61}$$

where $G_{Na_{max}}$ is the maximum conductance of the sodium channel, m represents the activation gate, h the fast inactivation gate and j the slow inactivation gate within the sodium channel. m, h and j each are described by their rate constants and time constants. V_{Na} is the reversal potential of sodium (see Eq. (2.51) in Section 2.3.1.1).

While earlier models like those of LUO AND RUDY [32, 33] and PRIEBE AND BEUCKELMANN [34] were based on animal data and described only major ionic currents, the TENTUSSCHER model [35] formulates a model based on the latest experimental data describing human ventricular behavior in much more detail. Ionic currents described in this model are:

I_{Na} : fast Na^+ inward current
I_{K1} : inward rectifier K^+ current
I_{to} : transient K^+ outward current
I_{Kr} : rapid, delayed K^+ outward current
I_{Ks} : slow, delayed K^+ outward current
I_{CaL} : L-type Ca^{2+} current

I_{NaCa} : Na^+/Ca^{2+} exchanger current
I_{NaK} : Na^+/K^+ pump current
I_{pCa} : Ca^{2+} plateau current
I_{pK} : K^+ plateau current
I_{bCa} : Ca^{2+} background current
I_{bNa} : K^+ background current

For threshold determination of action potential generation (Section 4.3), the reduced TENTUSSCHER model, based on these fundamental equations but still computationally cost efficient [36], was used in this thesis. Figure 2.7 visualizes the models' schemes of both TENTUSSCHER models.

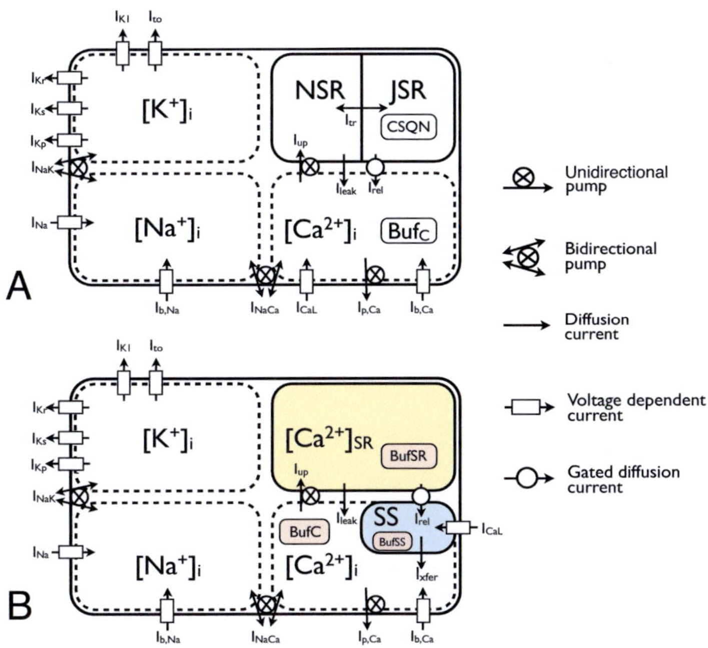

Fig. 2.7. Schemes of the TEN TUSSCHER human ventricular cell models [37]: (A) detailed model of 2004: Ion currents and compartments in the cytosol and the sarcoplasmatic reticulum (SR) with the corresponding calcium buffers. Bufc: cytosolic calcium buffers, NSR: network sarcoplasmic reticulum, JSR: junctional sarcoplasmic reticulum, CSQN: Calsequestrin. (B) reduced model, published by TEN TUSSCHER ET AL. in 2006. This model includes some simplifications for the benefit of computational cost reduction.

2.4 Effects of Time-Varying Magnetic Fields upon Biological Tissue

The term "effects of time-varying fields" encompasses both wanted and unwanted effects on biological tissue. "Wanted effects" refers to those used in therapeutical situations. Unwanted effects may be an unavoidable byproduct of a particular device for therapeutical field applications or effects to the general public who are not aware of being exposed in their everyday life. However, a biological effect does not automatically imply to be hazardous. This section describes the physical effects of low frequency magnetic fields on biological tissue without classification regarding health hazards.

Physical effects of EM fields are based on interactions with ions within tissue. As explained in Section 2.2.1, applied fields lead to motion of free charge, alignment of dipoles or the induction of dipoles. Depending on the field frequency, this leads to electric stimulation of excitable tissue or to heating due to energy absorption.

Magnetic fields by themselves do not have any effect to biological tissue. But time-varying magnetic fields induce electric fields which in turn affect biological tissue. The term "magnetic stimulation" usually refers to magnetically induced electric currents which cause nerve or muscle stimulation.

2.4.1 Stimulation of Excitable Tissue by Magnetic Fields

One characteristic of time-varying magnetic fields is the ability to stimulate excitable tissue. Due to their comparably low conductivity and geometry, peripheral nerves have the lowest stimulation threshold. Stimulation mechanisms apply also to neuronal tissue, skeletal muscle or myocardial tissue, but with increased stimulation thresholds as can be derived from the following stimulation mechanisms.

2.4.1.1 Mechanisms of Magnetic Stimulation

According to FARADAY's law, the **E** field is related to the time rate of change of the magnetic flux density **B** by

$$\oint \mathbf{E}\, d\mathbf{l} = -\frac{d}{dt}\int \mathbf{B}\, d\mathbf{a}. \tag{2.62}$$

$d\mathbf{l}$ denotes an element of a closed path $\mathbf{l}$, whereas $d\mathbf{a}$ is an area element. If $\mathbf{B}$ varies sinusoidally with uniform amplitude B_0 over the region inside $\mathbf{l}$ which might be a circle of radius r, the induced electric field strength E calculated from Eq. (2.62) is:

$$E = -\frac{r}{2}\frac{dB}{dt} = [r\pi f B_0]\cos(2\pi f t) \tag{2.63}$$

Within a volume conductor that is exposed to the **B** field, Eq. (2.63) implies that the outermost rings would have the greatest electric field strengths. If the **B** field is generated by a current loop, the highest field strengths will occur at the projection of that loop on the surface of the conductor.
The induced **E** field, as described above, sets free charges into motion, which is expressed by the current density. The relationship is:

$$J = \sigma E \tag{2.64}$$

with J being aligned with E.

2.4.1.2 Excitability of Neural Tissue

The excitability of tissue depends on several factors:

- stimulus waveform, magnitude and duration
- repetition pattern
- current distribution
- shape, location and state of the tissue

On a cellular level, the induced current meets a cell membrane and alters the membrane potential V_m. It has been shown, that in a homogeneous medium, a straight axon is activated at places where $\frac{\partial E_x}{\partial x}$ is maximum, with E_x being the component of the electric field along the axon. This implies that if the electric field was truly uniform, and the axon was infinite in both directions, there would in theory be zero net current transfer [38]. However, there are no infinite, straight axons (and internal fields are never uniform). Hence, excitation most probably takes place at bends of axons, at constrictions of nerve cells or at synaptic terminals [39, 40]. In other words, whenever the axon experiences a change in the electric field, the transmembrane voltage is altered.
At this point, an important difference between magnetic field stimulation and direct contact stimulation becomes evident: In contact stimulation, the current is injected into the extracellular space. This leads to large changes in the membrane potential in the area close to the electrode, since here the current has to be distributed between extra- and intracellular space. Magnetically induced electric fields, on the contrary, are evident in both extra- and intracellular space. Hence, the **E** field sets free charges into coherent motion in both media. At locations where the free mo-

tion is interrupted, the membrane gets either depolarized or hyperpolarized (see Fig. 2.8).

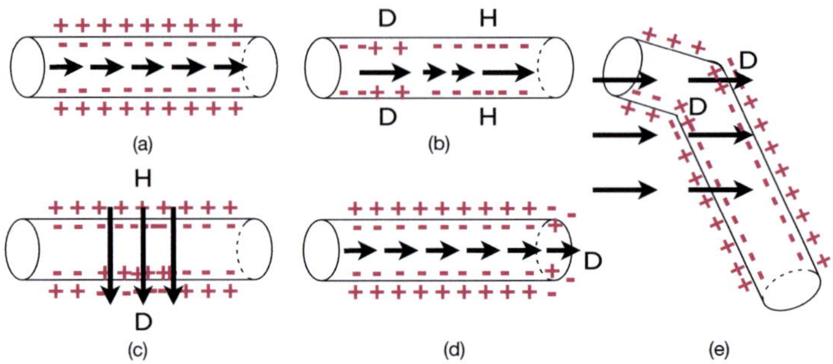

Fig. 2.8. Schematic illustrations of depolarization *(D)* and hyperpolarization *(H)* of the cell membrane, with identical **E** inside and outside of the cell: (a) uniform **E** field along the axon yields no change in the TMV. (b) $\partial E_x/\partial x \neq 0$ leads to cell activation. (c) transverse excitation, triggered by an **E** field locally across the axon. (d) axon terminating in a uniform **E** field. (e) bent axon in a uniform **E** field.

The subthreshold behavior of $\mathrm{V_m}$ has theoretically been modeled using the cable equation [38, 39, 41]

$$V_m + \tau\frac{\partial V_m}{\partial t} - \lambda^2\frac{\partial^2 V_m}{\partial x^2} = -f(x,t) = -\lambda^2\frac{\partial E_x}{\partial x} \tag{2.65}$$

with τ being the time constant of the membrane ($\tau = r_m c_m$, $r_m c_m$ being the membrane's capacity and resistivity, respectively), λ the length constant of the fiber and E_x the component of the electric field that is parallel to the fiber axis (x). The left-hand side is the conventional 1-D cable equation, the right-hand side ($f(x,t)$) is called the *activating function*. It shows that the driving force for nerve excitation is the second derivative of the applied potential field [38, 41].
Equation (2.65) has been developed for axons of infinite lengths in homogeneous fields, but also holds for bent axons or axons of finite lengths. Extended versions of the cable equation model also include axon dynamics, as formulated in the HODGKIN-HUXLEY model (see 2.3.2). With the cable equation model, subthreshold depolarization and hyperpolarization behavior of the cell membrane can be reproduced, for various characteristics of the **E** field.

2.4.1.3 Excitation by Mono-Phasic, Rectangular Stimuli: Strength-Duration Relation

An excitable cell is activated if the transmembrane voltage rises above the characteristic threshold potential. Hence, the **E** field needs to be strong enough in order to remove for example positive charge from the exterior cell membrane *(cathodal stimulus)*. Electric charge Q is the product of current flow I for the time t. This implies that a strong current applied for a short time has the same effect than a weaker current, applied for a longer time. This idea led to the strength-duration relation formulated by BLAIR in 1932 [42]:

$$I(t_p) = \frac{b}{1 - e^{-t_p/\tau}} \tag{2.66}$$

Here, t_p is the pulse duration of a rectangular stimulus and τ the membrane's time constant. b is the *rheobase* which denotes the minimal current amplitude of infinite duration. b and τ are characteristic tissue parameters and need to be investigated by measurement. Together with the rheobase, the term *chronaxie* was introduced. It denotes the time duration of an impulse of current strength twice the rheobase value that is long enough for stimulation. Figure 2.9 shows the relationship. Another strength-duration-relation was proposed even earlier by G. WEISS in 1901. The formulation was stated as [43,44]:

$$I(t_p) = I_0 \left(1 + \frac{\tau_e}{t_p}\right), \tag{2.67}$$

where I_0 is the minimum threshold current for long pulses (rheobase) and τ_e is the chronaxie.

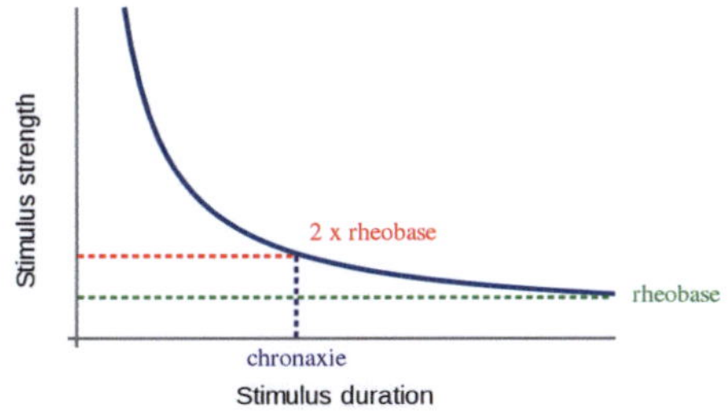

Fig. 2.9. Strength-duration curve including rheobase and chronaxie

Towards long pulse durations, V_m is altered very slowly, the membrane has time enough to reestablish its resting state and no activation will take place. On the other hand, if the stimulus is too short, the membrane's channels are not fast enough to react upon the event of potential increase, hence no sodium ions will enter the cell and no action potential will occur.

2.4.2 Thresholds for Peripheral Nerve and Muscle Stimulation of Time-Varying Magnetic Fields

In order to classify effects of time-varying magnetic fields, it is necessary to investigate critical magnitudes and distributions of induced fields and absorbed power within the body. Studies of direct contact stimulations of very low frequency currents (below 1 kHz) are manifold due to their widespread application (for example 50 Hz and 60 Hz power line frequency). On the other hand, power absorption caused by electromagnetic waves is investigated extensively, thanks to communication technology and magnetic resonance techniques, to mention only two of them.

2.4.2.1 Theoretical Models

The frequency range from 1 kHz to 100 kHz has not been investigated intensively yet. Exposure to magnetic fields of that frequency range may include unpleasant or painful sensations, involuntary muscle contractions, burns, cardiac arrhythmias and cardiac fibrillation [38]. The strength-duration relation (Section 2.4.1.3) gives an estimation of the stimulation thresholds of direct contact currents. For magnetic field stimulation, J. PATRICK REILLY created a model for threshold estimation of cardiac, skeletal muscle and neural tissue. His work is based on a model that incorporates an extended version of the cable equation (2.65) which he introduced as the *spatially extended nonlinear node (SENN) model* [38,45]. From this model, he derived basic stimulation thresholds for monophasic rectangular stimuli of $t \geq 1$ ms and $t \leq 10\,\mu$s, which are listed in Table 2.3.
The magnetic field exposure thresholds determined by REILLY have been based on a number of conservative assumptions. For example, the exposure area has been assumed to be 0.2 m, penetrated by a spatially uniform field.
From his findings and a simplified body model, REILLY defined strength-frequency relations for stimulating **E** fields, induced due to magnetic field exposure. This strength-frequency relations are defined by two parameters: E_0 and τ_e, where E_0 is the minimum excitation threshold with long duration stimuli (rheobase), and

	Criteria		Fiber diameter [μm] 5	10	20
E	[V/m]	($t_p \geq 1$ ms)	24.6	12.3	6.2
$E \cdot t_p$	[Vs/m]	($t_p \leq 10\,\mu$s)	2.98e-03	1.49e-03	7.5e-04
J	[A/m^2]	($t_p \geq 1$ ms)	4.92	2.46	1.23
q	[C/m^2]	($t_p \leq 10\,\mu$s)	6.0e-04	3.0e-04	1.5e-04
dB/dt	[T/s]	($t_p \geq 1$ ms)	246	123	62
$dB/dt \cdot t_p$	[mT]	($t_p \leq 10\,\mu$s)	30	15	7.5

Table 2.3. Minimum stimulation thresholds with uniform field excitation; single monophasic stimuli [38]

τ_e is the equivalent membrane time constant, which defines the stimulus duration below which monophasic stimulus thresholds converge to a line that is inversely proportional to stimulus duration [46]. Values of E_0 and τ_e depend on tissue type and stimulus waveform. According to REILLY, the relationship is the following:

$$E_i = \begin{cases} E_0 & \text{if } f \leq f_e, \\ E_0 \frac{f}{f_e} & \text{if } f \geq f_e, \end{cases} \tag{2.68}$$

with $f_e = 1/(2\tau_e)$ [46,47]. E_0, τ_e and f_e for nerve and cardiac excitation, are given in Table 2.4.

Reaction	E_0 [V/m]	$B_{0,\text{eff}}$ [mT]	dB/dt [T/s]	τ_e [ms]	f_e [Hz]
Synapse activity alteration, brain	0.075	8.14	1.45	25.0	20
10-μm nerve excitation, brain	12.30	7.97	2.37	0.149	3,350
20-μm nerve excitation, body	6.15	1.27	37.5	0.149	3,350
Cardiac excitation	12.0	59.8	88.7	3.0	167

Table 2.4. Parameters for threshold models of reaction. The exposure radius r (Eq. (2.63)) is assumed to be 0.2 m [46]

The group of IRNICH states that the best formula for describing electrostimulation by time-varying magnetic fields is given by

$$\int E dt \geq E_{rheo} t_{chron} \left(1 + \frac{t_p}{t_{\text{chron}}}\right), \tag{2.69}$$

where t_p is the duration of the electric field E, E_{rheo} is the rheobase of the electric field, and t_{chron} is the chronaxie time of the physiologic system [48]. This provides another formulation of the strength-duration relation.

2.4.2.2 Experimental Data

Experimental data upon peripheral nerve and muscle stimulation do not always provide a clear description of all the relevant details, such as a complete description of the stimulus waveform, its spatial orientation, the method of exposure, the biological preparation or subject, or the criterion for judging the presence of excitation. Experiments upon functionality of nerve and muscle stimulation with contact electrodes have started in the 1960's. Groups like those of WYSS ET AL. [49, 50], BROMM ET AL. [51], CHATTERJEE ET AL. [52] and GEDDES ET AL. [53] studied the characteristics of stimulation by sinusoidal contact currents at different frequencies. Table 2.5 lists an extract of available data of magnetic field exposure studies.

Stimulus waveform	t_p [ms]	Subject/ Preparation	Body Focus	Response	threshold peak E [V/m]	J [A/m^2]	Ref.
damped cosine 3 kHz	2.0	rat	chest	twitch	34	5.1	[54]
1/2 cycle cosine	0.083	rat	hest	twitch	45	6.8	[54]
1/2 cycle cosine	0.15	rat	hest	twitch	36	5.4	[54]
damped cosine 3 kHz	2.0	human	forearm	sensation	25–50	3.8–7.5	[54]
damped cosine 3 kHz	–	human	forearm	EMG	78	11.8	[54]
1/2 cycle cosine	0.12	human	forearm	EMG	113	16.7	[54]
1/2 cycle cosine	0.24	human	forearm	EMG	87	13.6	[54]
monophasic pulse	0.18	rat	whole	muscle	44–73	9–16	[55]
monophasic pulse	018	human	wrist	EMG	70	14	[56]
pulse	2–3	frog nerve			64–96	14–32	[57]
1/2 cycle cosine	2.0	frog nerve			15	3.0	[58]

Table 2.5. Experimental thresholds for neural stimulation via exposure to time-varying magnetic fields. Information is partly incomplete.

It is difficult to compare experimental data against each other, since experimental setups vary quite much and information is partly incomplete.

2.4.3 Energy Absorption and Temperature Increase

Besides electric stimulation, there is one other physical effect of time-varying magnetic fields to biological tissue. In fact, it is again not the magnetic field that affects the tissue but the induced electric field that transfers energy to electric charges within tissue by exerting forces to them. The force, that a **B** exerts on the charges can change their directions but not their energy. The **B** field transfers energy through forces on permanent magnetic dipoles. However, biological tissue has only very few permanent magnetic dipoles, hence this effect is negligible.

2.4.3.1 Specific Absorption Rate

The complex valued electric power, which is the time rate of energy, transferred to charges in an infinitesimal volume element ΔV of a material is given by

$$P_E = \frac{1}{2}\overline{\mathbf{J}} \cdot \mathbf{E} \cdot \Delta V \tag{2.70}$$

with $\overline{\mathbf{J}}$ being the conjugate of the complex valued current density. Substitution of $\overline{\mathbf{J}}$ by $\overline{\sigma(\omega)\mathbf{E}}$ results in

$$P_E = \frac{1}{2}\overline{\sigma(\omega)} \cdot |\mathbf{E}|^2 \cdot \Delta V \tag{2.71}$$

The real part of Eq. (2.71) is determined by the effective conductivity and yields the effective power

$$P_{\mathit{eff}} = \frac{1}{2}\sigma_{\mathit{eff}}(\omega) \cdot |\mathbf{E}|^2 \cdot \Delta V \tag{2.72}$$

Usually, the absorbed power is normalized with the volume mass, which leads to the definition of the *specific absorption rate (SAR)*:

$$SAR = \sigma_{\mathit{eff}}(\omega) \cdot \frac{|\mathbf{E}|^2}{2\rho} \tag{2.73}$$

The SAR provides information of the time rate of energy at the specific point where $|\mathbf{E}|$ is evident. Hence the relation given in (2.73) is referred to as *local SAR*. Power absorption depends on tissue characteristics and rises with frequency of the magnetic field, since $\mathbf{E}$ is frequency dependent.

Power absorption is always related to thermal heating. Ignoring all thermoregulatory systems, temperature increase by power absorption can be estimated by multiplying the SAR with the interaction time of tissue with the magnetic field t_{WW} and dividing by the tissue's specific heat capacity C:

$$\Delta T = \frac{SAR \cdot t_{WW}}{C} = \cdot \frac{\sigma_{\mathit{eff}}(\omega)|\mathbf{E}|^2 \cdot t_{WW}}{2\rho C} \tag{2.74}$$

However, considering tissue warming in the human body for example, this provides an unrealistic estimation of temperature increase since there are several mechanisms of heat transfer, which will be described in the following section.

2.4.3.2 Thermoregulatory System of the Human Body

The natural core body temperature of humans is kept constant at $\approx 37°C$ by the body's thermoregulatory system. Peripheral temperature and the temperature in the limbs vary with surrounding temperature and activity between 28°C and 37°C. The sympathetic nervous system directs the body's attempts to regulate temperature by triggering heat loss or heat production in case of core heating or cooling, respectively. Heat production is achieved by

- metabolic heat produced by chemical reactions
- muscular activity
- increase of metabolic process triggered by the release of stress hormones of the sympathetic nervous system

On the contrary, heat loss is realized by

- radiation to the surrounding
- conduction: heat transfer by direct contact, molecule to molecule
- convection: movement of molecules along the spatial temperature gradient
- evaporation of sweat

In case of local warming, convection and conduction are responsible for transferring heat to cooler areas. Increasing the diameter of the capillaries *(vasodilation)* increases the cooling effect by rising blood flow.

2.4.3.3 Calculation of Temperature Increase: The PENNES Bio-Heat Equation

Temperature distribution in body tissues is most commonly described using PENNES' *bio-heat equation.* It combines the heat flow theory (FICK principle) in basic terms of the local rate of tissue heat production and volume flow of blood and the NEWTON cooling law applied to the heat transfer between skin and environment [59]. Heat transfer mechanisms of blood can be formulated as:

$$h_b = \omega_b \rho_b C_b \rho (T - T_b) \tag{2.75}$$

where h_b is the rate of heat transfer per unit volume of tissue, ω_b is the perfusion rate per unit volume of tissue, ρ_b is the density of blood, C_b is the specific heat of blood, ρ the density of tissue, T_b is the blood temperature, and T is the local tissue temperature.

The thermal energy balance for perfused tissue is expressed in the following form, along with thermodynamic theory:

$$\rho C \frac{\partial T}{\partial t} = k\nabla^2 T + \rho Q_m - h_b \tag{2.76}$$

where C is the specific heat capacity, k is the thermal conductivity of tissue, and Q_m is the rate of metabolic heat production rate.
Up to this point, the formulation accounts for metabolic heating, heat conduction and heat transfer by perfusion. The absorbed power contributes as an additional heat source to the temperature increase:

$$\rho C \frac{\partial T}{\partial t} = \nabla \cdot (k\nabla T) + \rho Q + \rho SAR - h_b \tag{2.77}$$

Cooling mechanism of convection is not included in the formulation. Hence, calculations of temperature increase always underestimates tissue cooling.

2.4.4 Exposure Guidelines for Limiting Effects of Time-Varying Magnetic Fields

Guidelines for the limitation of exposure to electric and magnetic fields are based on recognized and reproducible interactions between the fields and the body. These effects were all acute effects of exposure to EM fields on excitable tissue, such as nerve and muscle stimulation or thermal heating due to power deposit. The evidence for chronic effects has until today not been considered sufficiently robust to justify setting exposure limits limiting those.
Guidelines of various organizations are designed to protect against direct and indirect effects of field exposure on body tissues for a wide frequency range. Effects include effects on biological tissue and adverse health effects. An adverse health effect causes detectable impairment of the health of the exposed individual or of his or her offspring; a biological effect, on the other hand, may or may not result in an adverse health effect [60]. Those include perception and annoyance through surface electric charge effects and the stimulation of central and peripheral nervous tissues. The induction in the retina of phosphenes, a perception of faint flickering light in the periphery of the visual field, is the most sensitive effect of field exposure and therefore accounted for in the design of exposure guidelines.
In 1992, the International Commission on Non-Ionizing Radiation Protection (ICNIRP) was established as a successor of the International Non-Ionizing Radiation

Committee (INIRC) which was a group formed by the International Radiation Protection Association (IRPA) in 1974.
Existing guidelines are based on review of literature on biological effects and an evaluation of the health risks of exposure to non-ionizing radiation. These health criteria have provided the scientific database for the subsequent development of exposure limits and codes of practice relating to non-ionizing radiation. These guidelines are periodically revised and updated as advances are made in identifying the adverse health effects of time-varying EM fields [60].
Exposure limits cited in this thesis always refer to those published by the ICNIRP. They presented two classes of guidance:

Basic restrictions: In the guidelines of 1998, restrictions on exposure to time-varying EM fields that are based directly on established health effects are referred to as *basic restrictions*. Depending on the field frequency, the physical quantities used to specify these restrictions are current density, specific absorption rate (SAR), and power density [60].

Reference levels: Since basic restriction quantities may be hard to assess, the reference levels are provided for practical exposure assessment purposes. Some reference levels are derived from relevant basic restrictions using measurement and/or computational techniques, and some address perception and adverse indirect effects of exposure to EMF. The derived quantities are for example electric field strength, magnetic field strength, magnetic flux density, power density and currents flowing through the limbs. If the measured or calculated value exceeds the reference level, it does not necessarily follow that the basic restriction will be exceeded. However, whenever a reference level is exceeded it is necessary to test compliance with the relevant basic restriction and to determine whether additional protective measures are necessary [60]

In 2010, an update of the guidelines from 1998 was published, replacing the low frequency part of the earlier version. Biological effects of exposure to low frequency EM fields have been reviewed by the International Agency for Research on Cancer (IARC), ICNIRP, and the World Health Organization (WHO) and national expert groups. Those publications provided the scientific basis for these guidelines [61]. In the new guidelines, the current density is no longer considered a basic restriction. It was replaced by the internal electric field, referring to the result within the body of electric or magnetic field exposure.
The reference levels are obtained from the basic restrictions by mathematical modeling using published data [61]. They were calculated for the condition of maxi-

mum coupling of the field to the exposed individual, thereby providing maximum protection. The reference levels consider induced electric fields in the central nervous system (CNS) and in non-CNS tissues anywhere in the body,

Guidelines for basic restrictions and reference levels differentiate between limits for "occupational exposure" (referring to non-accidental time-limited exposure) and restrictions for the general public. Limits for occupational exposure include a safety factor of 10 to established threshold data. This accounts for the statistical characteristic of experimentally obtained results. Limits for the general public contain an additional safety factor of 5.

The SAR is not included into low frequency exposure guidelines. This is because thermal heating is not regarded as a serious issue in field exposure to low frequency fields. As a rule of thumb, it is accepted to tolerate a temperature increase of 1 K within the course of 1 hour field exposure. SAR restrictions which ensure compliance with that, are extracted from guidelines for Magnetic Resonance Imaging [62]. The relevant data of basic restrictions, reference levels and MR guidelines are summarized in Table 2.6.

Basic restriction ICNIRP 1998	J_{eff}	1 – 100 kHz	f/100 A/m^2
Basic restriction ICNIRP 2010	$E_{eff,internal}$	3 kHz – 10 MHz	0.27 · f V/m
Reference level ICNIRP 1998	B_{eff}	0.82 — 65 kHz	3.07e-05 T
		65 kHz – 1 MHz	2.0e-03 /f T
Reference level ICNIRP 2010	B_{eff}	3 kHz – 10 MHz	1.0e-04 T
INIRC	whole body SAR		1 W/kg for 1 h
	SAR × time		120 W · min/kg
	SAR instantaneous		8 W/kg
	SAR_{10g}		10 W/kg

Table 2.6. Summarized restrictions for occupational exposure (ICNIRP) [60, 61] and magnetic resonance examination (INIRC) [62]. f is in kHz, effective values refer to the $1/\sqrt{2}$ amplitude.

Part II

Investigating Field Effects upon Excitable Tissue

3

Human Body Simulations within the Scope of the MAGIC Project

3.1 Introduction

The MAGIC joint research project was initiated by PHILIPS Research in Hamburg, Germany. Supported by the German Ministry of Education and Research (BMBF), many work packages have been formulated. The IBT was assigned with the project presented in this chapter, which became part of this thesis.
The following sections deal with the field effects induced by the field generating coils of the MPI system. Coil specifications are provided by PHILIPS, including coil geometry and target central magnetic flux density, in terms of amplitude and frequency. The existing MPI system by the time this thesis was developed was a mouse scanner for laboratory research usage only. Time-varying fields are generated by the drive field coils. They produce a magnetic flux density in the center of the field of view of 20 mT amplitude at frequencies around 25 kHz. For the following considerations, the coil geometry was scaled by a factor of 10, so that a human body model fits into the scanner. The target **B** field amplitude was set to 10 mT, from which quantities can be scaled easily to whatever might be the final value in future MPI systems. Simulations have been carried out at frequencies up to 100 kHz. However, most attention is paid to results of 25 kHz simulations.

3.2 Electro-Magnetic Simulations on a Human Body Model

Concluding from previous chapters, quantities that have to be paid attention to when exposing patients to time-varying magnetic fields of the low kHz range for imaging purposes, are current density J, that might induce nerve or muscle stimulation, and specific absorption rate *SAR* that is a measure for thermal heating. Using numerical methods, these quantities can be calculated at every point in a human model. This section presents the methods, resources and results of the field calculations on a human body model in MPI drive field coils.

3.2.1 Setting up the Simulation

The numerical simulations require a voxel model, representing a human patient, a model of the MPI system, dielectric tissue parameters and MPI settings. The resources of these ingredients will be described in the following sections. To start with, the chosen field calculation software is introduced.

3.2.1.1 *SEMCAD X* Software Package

SEMCAD X [22] is a 3-D full wave simulation environment, developed and provided by SCHMID & PARTNER ENGINEERING (SPEAG) in Zürich, Switzerland. Among others, *SEMCAD X* provides standard radio frequency solvers as well as solvers for low frequency problems and solutions for coupled EM-thermal simulations. Furthermore, a range of specific method enhancements have been integrated. The 3-D solid modeler allows the rapid import and processing of various CAD *(Computer Added Design)* formats and features a fast OGL *(Open Graphics Library)* based rendering engine. The post processor provides extraction of any EM, SAR or temperature related result and its fast 3-D rendering within the graphical user interface (GUI). A python interpreter enables the user to automize all processes from geometry modeling, simulation setting and post processing. This simplifies and shortens procedures when many simulations are on due with only minor changes in the settings.
Running a simulation in *SEMCAD X* takes several steps from modeling, simulation setting and post processing.

Modeling: In modeling mode, one can create and modify solids, sources, lumped elements and sensors in the model. Modifications are carried out by rotating, scaling, translating single or grouped parts, combining, extruding and many more operations.

Simulation setting: In simulation mode, all required settings are made. It is recommended to process through the settings step by step. First, the solver type, simulation frequency and the solver's convergence tolerance are determined. Next, all solid regions need to be assigned their characteristic properties depending on their solid type. For dielectric materials, electric conductivity in S/m and relative permittivity are relevant specifications. Sources can be assigned amplitude and phase. Boundary conditions can be VON NEUMANN or

DIRICHLET (see Section 2.2.3) conditions. Finally, settings for the computational grid must be made. Since the accuracy of simulation results depend highly on grid quality, much attention is to spend on generating a geometrically accurate numerical representation of the model. The grid generator creates a non-uniform rectilinear grid, which automatically adapts to the details of the model. From there, the user decides about priority of necessary accuracy. The user is informed about grid statistics and number of lines instantaneously a change was made. At the end, the "voxeler" is started which is the last step before the simulation is ready to run. Generation of the mesh containing 45 million nodes takes about an hour on a 2.2 GHz Dual Core AMD Opteron Processor with 16 GB RAM. In total, one EM simulation takes about 16 hours.

Post processing: The *SEMCAD X* post processing engine provides enhanced data extraction and handling as well as visualization capabilities. Result files are platform independent, so a simulation that was run on a Linux or Windows (32 or 64 bit) machine may be viewed by using any 32 or 64 bit Windows computer with *SEMCAD X* installed. All field quantities are accessible via the respective sensor recording those. Field matrices can be exported for further handling in *Matlab* or as plane text files.

3.2.1.2 Visible Human Data Set

The model that was chosen for the simulations originates from the Visible Human Project, from the National LIbrary of Medicine, Bethesda, USA [63] [64]. Back in the 1970's, the corps of an executed convict was deep frozen, sliced and photographed [65]. From these pictures 44 tissue types had later been segmented and classified at the Institute of Biomedical Engineering, Universität Karlsruhe (TH). The resulting voxel model is available with 1 mm resolution and 2 mm resolution. For field calculations, 2 mm resolution is more than sufficient. Finer structures can not be accounted for due to limitations of the computational grid. To further reduce model size, only the torso of the data set has been used in the field simulations, containing 129 million cubic voxels. Since the segmented model did not include skin tissue, the model was expanded by one voxel layer, representing a thin layer of skin tissue. This simplification does not account for real skin and makes generation of the computational grid even more difficult. Keeping the number of computational nodes down at a reasonable number and at the same time modeling a closed skin surface is impossible. The grid has been refined until the maximum number of voxels was reached. Resolution reaches 1 mm in detailed regions and does not exceed 5 mm within the entire model. Figure 3.1 presents the intermediate slice of the segmented model.

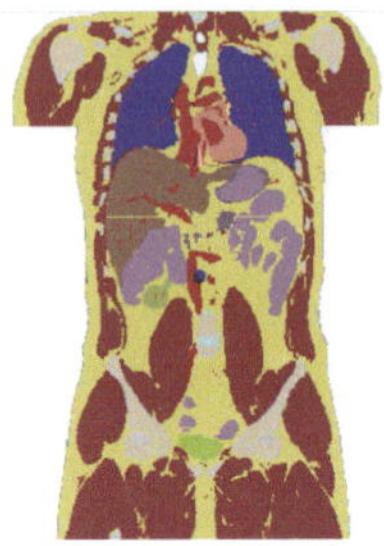

Fig. 3.1. Intermediate slice of the segmented voxel model of the Visible Human

The torso of the body model contains 17 tissue types. They have all been assigned with dielectric properties published by GABRIEL ET AL. [28, 66, 67]. Table 3.1 lists all tissue types in the model together with respective characteristic values for conductivity and relative permittivity at 25 kHz, necessary for current density and SAR calculations, and thermal conductivity, specific heat capacity and perfusion rates, necessary for estimations of temperature increase.
The entry “Intestines” refers to the combination of small and large intestines. These organs are not separately classified in the Visible Human model. However the conductivity of both tissue types differ by a factor of around 40 according to GABRIEL ET AL. Thus, a trade-off has been found that averages dielectric properties as it is described in [68].

3.2.1.3 Modeling MPI Coils

The MPI system consists of two sets of field generating coils: the selection field coils that generate a constant field gradient and thus the field free point (FFP). The second coil set consists of three coil pairs, generating the drive field and focus field. The drive field moves the FFP along a Lissajous trajectory through the bounded field of view, whereas the focus field moves the field of view successively (Section 2.1). Every coil pair generates a homogeneous magnetic field distribution oriented parallel to the coil pair’s axis. Regarding geometry, there are numerous possibilities to realize a homogenous field. The simplest one is the HELMHOLTZ geometry. The field generating coils, from here on referred to as *x-coils, y-coils*

	σ [S/m]	ε_r	C [$J/kg/K$]	k [$W/m/K$]	ρ [kg/m^3]	ω [$ml/min/kg$]	Q [W/kg]
Bladder	0.215	3177	3900	0.56	1040	78	1.54
Blood	0.700	5228	3824	0.51	1060	10000	-
Bone	0.083	841	1613	0.39	1990	30	0.30
Bone marrow	0.103	360	4191	0.60	1027	-	5.55
Cartilage	0.176	3025	3664	0.47	1100	50	1.46
CSF	2.000	109	4191	0.60	1007	-	-
Fat	0.043	287	2524	0.25	916	27	0.33
Heart	0.178	28834	3720	0.54	1060	900	9.06
Intestines	0.442	15774	3653	0.56	1044	1000	0.00
Kidney	0.150	17531	3745	0.52	1046	2612	45.89
Liver	0.063	15356	3600	0.51	1050	1007	11.43
Lung	0.254	14097	3625	0.44	655	400	2.60
Muscle	0.346	13172	3546	0.53	1041	28	0.46
Nerve	0.058	16838	3664	0.46	1038	549	6.84
Skin (dry)	0.000	1131	3437	0.35	1100	97	1.47
Spleen	0.115	7444	3603	0.54	1054	1142	14.32
Stomach	0.532	4673	3553	0.53	1050	374	4.95

Table 3.1. Dielectric and thermal properties assigned to torso model tissues: Quantities are: σ: electrical conductivity, ε_r: relative permittivity, C: specific heat capacity, k: thermal conductivity, ρ: tissue density, ω: perfusion rate, Q: metabolic heat generation rate. Dielectric values are given by the Gabriel parametric model for body tissues at 25 kHz [28]. All values are drawn from the material database of *SEMCAD X*

and *z-coils* with respect to the orientation of the generated **B** field, are modeled by single virtual current sources, enclosing the volume of interest. The shape parameters have been provided by PHILIPS, representing the coil realization of the latest mouse scanner. For the human simulation setup, the mouse geometry has been scaled up by a factor of 10, resulting in a bore hole diameter of 1.2 m. Fig. 3.2 illustrates the setup. Frontally and sagittally oriented coil pairs are saddle coils, spanning an imaginary cylindrical tube. Apart from boundary regions, the saddle coils generate a widely homogeneous magnetic field similar to HELMHOLTZ coils characteristics. The amplitude of the magnetic flux density vector **B** can be calculated using the BIOT-SAVART-law, which provides **B** at any point **r** induced by a given current density distribution (Section 2.2.1):

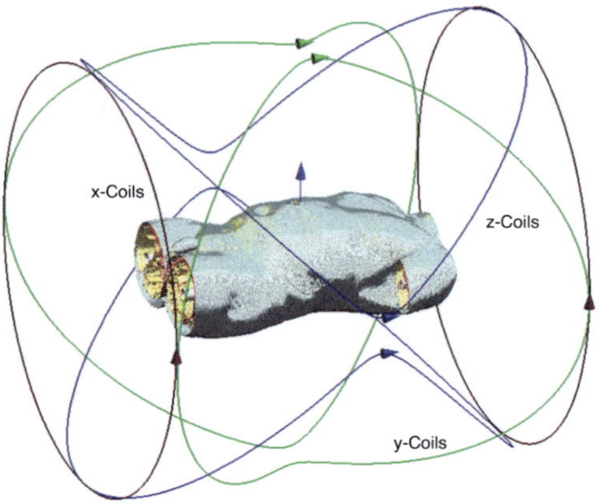

Fig. 3.2. Simulation model including the Visible Human torso and MPI drive field coils. The torso contains 17 tissue types. The dimensions measure 1.2 m in x-, y- and z-direction, counting $448 \times 384 \times 263$ grid lines, totaling in 45.24 million grid cells.

$$\mathbf{B}(\mathbf{r}) = \frac{\mu_0}{4\pi} \int_V \mathbf{J}(\mathbf{r}) \times \frac{\mathbf{r} - \mathbf{r}'}{|\mathbf{r} - \mathbf{r}'|^3} \tag{3.1}$$

Regarding the **B** field on the symmetry axis of one current loop of radius r, current I and N turns, Equation (3.1) simplifies to the following:

$$\mathbf{B}(x) = \frac{\mu_0 IN}{2} \cdot \frac{r^2}{(r^2 + x^2)^{3/2}} \cdot \mathbf{e}_x \tag{3.2}$$

Considering two current loops of distance R and translating the origin of the coordinate system to the center of the coil pair, the resulting central **B** becomes

$$\mathbf{B}(r/2) + \mathbf{B}(-r/2) = \mu_0 \cdot \frac{INR^2}{(5/4r^2)^{3/2}} = \mu_0 \cdot \frac{8IN}{\sqrt{125}r}. \tag{3.3}$$

In order to generate a central **B** field amplitude of 10 mT, which is the target flux density of all following simulations, a current strength of 6,672 Ampere-turns in each coil is necessary. Fig. 3.3 shows profiles of the generated **B** field of each coil pair.

Regarding field effects of time-varying magnetic fields, current densities (J) and specific absorption rates *(SAR)* are of interest (Section 2.4). In order to regard drive field effects, it is assumed that all coil components are driven at the same frequency. Thus the superposition principle is applicable so that all quantities of

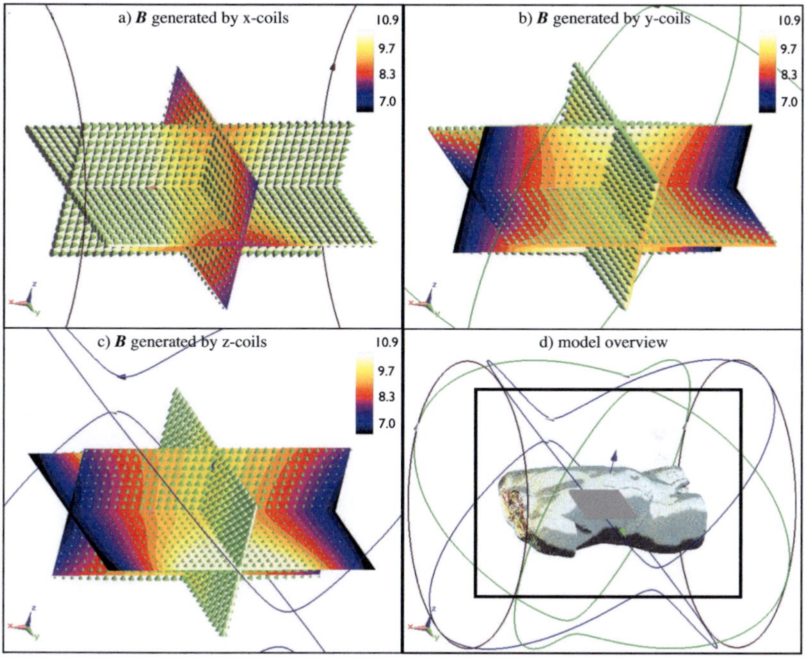

Fig. 3.3. Slice field views of **B** field vectors, generated by each coil pair separately. Field vectors are displayed, within the fields of view in which the torso is placed. **d** gives an overview of model orientation. The black frame depicts the display section.

interest can be derived from three single simulations, one for each coil pair. The case of differing frequencies, necessary for moving the FFP, will be regarded in section 3.2.4.

3.2.2 Drive Field Induced Fields

According to FARADAY's law, the induced electric field can be estimated to be proportional to the radius of the contemplated object, and to the frequency of the applied field. This originates from Maxwell's equation:

$$\nabla \times \mathbf{E} = -\dot{B} \tag{3.4}$$

In case of circularly oriented electric fields (straightly oriented magnetic flux) and sinusoidally oscillating **B** fields, Eq. 3.4 goes to

$$\mathbf{E} = \pi r f B_0 \cos \omega t. \tag{3.5}$$

Here, the linear relations between **E** and r and f and **B** become clear. The quantities of highest interest regarding biological effects are the induced current density and specific absorption rate. The relation between E, J and SAR is

$$\mathbf{J} = \sigma \cdot \mathbf{E} \tag{3.6}$$

and

$$SAR = \frac{\sigma}{\rho} E^2. \tag{3.7}$$

The destined drive field frequency for Magnetic Particle Imaging was 25 kHz at the time this thesis was developed. In order to cover the frequency range up to 100 kHz, several scenarios at different frequencies have been simulated. As can be derived from Eq. (3.6) and (3.7), current densities increase linearly with the applied frequency, whereas SAR exhibits quadratic growth.
In the following, simulation results will be presented referring to field excitation with an oscillating magnetic flux density of 10 mT amplitude and 10 kHz, 25 kHz, 50 kHz and 100 kHz frequency. Each coil pair is observed separately. Coil pairs are referred to according to the field direction they generate.

3.2.2.1 Induced Current Density

The current density vector **J** is proportional to the induced electric field with the tissue's conductivity as proportionality factor: $\mathbf{J} = \sigma \cdot \mathbf{E}$. Figure 3.4 illustrates current densities distributions in the x-, y- and z-plane of the setup. The maximum amplitude reached is 187 A/m^2. Depending on the **B** field orientation, current loops circulate on a larger radius, which leads to high values at the body's periphery. Due to its high conductivity, maximum values are reached in peripheral muscle tissue.
Detailed information about current distributions are provided in Fig. 3.5. Maximum and averaged current density amplitudes induced by different **B** field frequencies and coils are plotted with respect to each body tissue. One recognizes the linear correspondence between current density and **B** field frequency. The effect of increasing electric conductivity with frequency, which also leads to higher current densities ($\mathbf{J} = \sigma(f) \cdot \mathbf{E}(\mathbf{f})$), is much weaker than the rise of electric field strength with frequency. Mean values refer to tissue volume averages. Values are high in

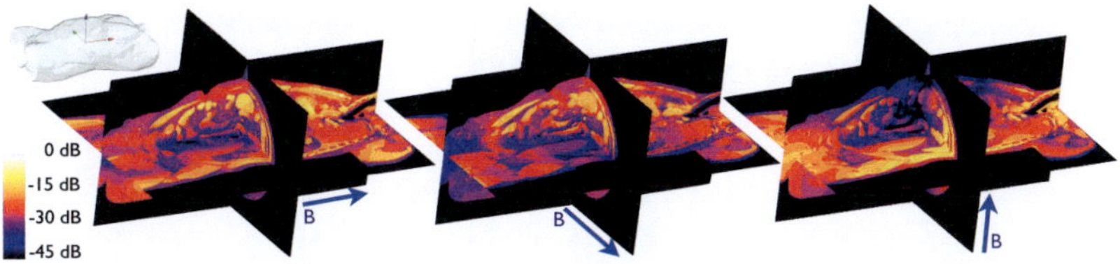

Fig. 3.4. Qualitative current distributions within the lying torso, induced by one coil pair (10 mT, 25 kHz), respectively. Depending on **B** field orientation, current loops circulate on a larger radius and thus achieve higher values. The maximum amplitude reached here is 187 A/m^2.

case of high electric conductivity or peripheral location. Both conditions are fulfilled in peripheral muscle tissue. Considerably high values also occur in blood, cerebrospinal fluid (CSF) and inner organs. However, currents are without effect there. Critical organs are the heart and skeletal muscle. Table 3.2 lists critical current amplitudes in myocardial and skeletal muscle tissue. More details on induced currents are given in the Appendix. There, cumulative histograms of J amplitudes in critical organs are given for every simulated frequency.

		J_{max}[A/m^2]				J_{mean} [A/m^2]			
		10 kHz	25 kHz	50 kHz	100 kHz	10 kHz	25 kHz	50 kHz	100 kHz
heart	x-coils	14.3	41.4	80.9	171.3	3.6	10.9	21.5	46.1
	y-coils	23.6	63.5	132.2	280.8	2.9	8.1	17.3	37.4
	z-coils	10.6	28.5	62.3	134.4	1.9	5.2	11.5	25.2
muscle	x-coils	54.9	151.5	288.6	602.5	8.0	21.9	41.3	84.7
	y-coils	49.2	125.6	251.8	517.3	7.3	18.7	37.7	77.4
	z-coils	67.0	168.0	351.0	732.0	9.5	23.7	49.1	100.8

Table 3.2. Maximum and averaged induced current densities in critical tissues heart and skeletal muscle. Peaks are very locally limited values, averages are statistical means over tissue volume.

At this point, there is no chance to make an assumption upon excitation of the heart, causing extrasystoles or even an arrhythmia, or upon stimulation of muscle tissue, causing a tickle or an even stronger penetration in the patient's body. Maximum values shown here are very much localized, occurring in few voxels only. Whether this is enough to cause an action potential, which is the origin of all muscular activity, will be discussed in Chapter 4.

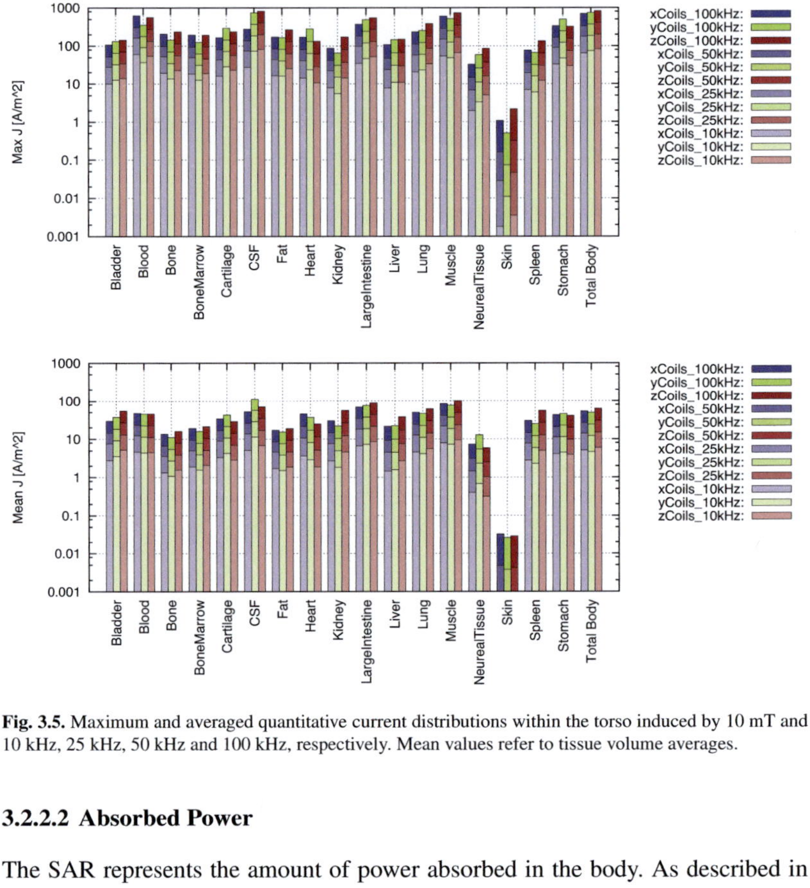

Fig. 3.5. Maximum and averaged quantitative current distributions within the torso induced by 10 mT and 10 kHz, 25 kHz, 50 kHz and 100 kHz, respectively. Mean values refer to tissue volume averages.

3.2.2.2 Absorbed Power

The SAR represents the amount of power absorbed in the body. As described in Section 2.4, SAR is commonly used as a measure for body warming. With regard to temperature increase, SAR usually is averaged over 10 g body tissue. When the term "SAR" is mentioned here, it is referred to the 10 g averaged value. *SEMCAD X* offers averaged SAR extraction following the IEEE 2002 standard [69].

Within MPI drive field settings, the applied field power grows quadratically with field frequency. Like current density, SAR originates from the induced electric field. Thus it is largest where circular loops of the electric field are big. Hot spots occur where the power cannot be transferred to neighboring tissues, thus in superficial areas. Fig. 3.6 illustrates qualitative SAR distributions, evoked by x-, y- and z-coil pairs, respectively. Maximum values at 25 kHz, 10 mT **B** field amplitude reach 8.15 W/kg generated by the x-coils, 6.04 W/kg by y-coils and 12.1 W/kg evoked by the z-coils. Outside those hot spots, the SAR is distributed evenly in wide areas of the body, being around 2 – 3 W/kg. Inner organs are hardly affected.

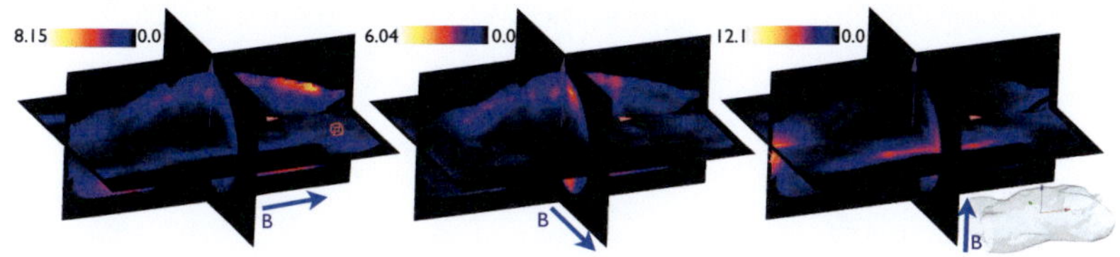

Fig. 3.6. Qualitative SAR distributions within the lying torso, induced by one coil pair, respectively. Evoked maximum values differ depending on the **B** field direction between 6 W/kg and 12 W/kg.

Analog to the current density bar plot (Fig. 3.5), Fig. 3.7 informs about maximum and statistical means of spatial averaged SAR of all torso tissues up to $f = 100$ kHz. Both local peaks and averaged values exceed MRI exposure limits (see Section 2.4.4).

Skin, which is a very poor conductor, shows the smallest induced currents and SAR despite its peripheral location. Muscle and fat tissues presents the highest values.

3.2.2.3 Estimated Temperature Increase

SEMCAD's thermal solver solves a Poisson differential equation considering a set of flexible boundary conditions. A source term describes the influence of tissue perfusion and of metabolic processes in the body. Accounting for non-linearities disables the usage of "Fast Poisson Solvers" (based on Fourier Transformation of the differential equation) and linear matrix-inversion-factorization methods. Furthermore, the evolution of heating over time suggests using a stepwise integration method for solving the equations. Therefore *SEMCAD X* resorts to the finite differences time domain (FDTD) method for solving the problem by using a variant on a non-uniform grid [22].

For temperature increase calculations, the SAR as result of the electro-magnetic (EM) simulation can be used as thermal source. Thus the equation that needs to be solved by the thermal solver, is the extended PENNES-equation:

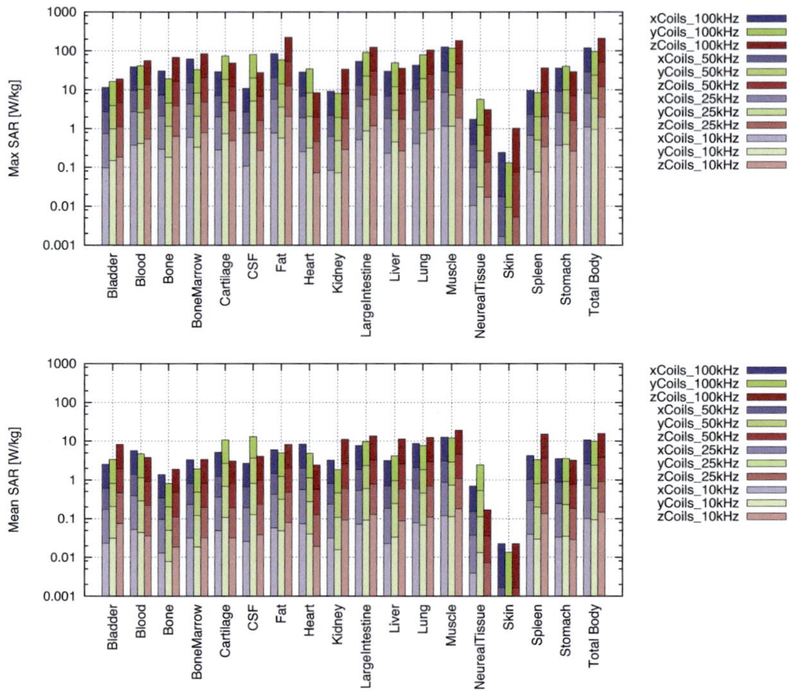

Fig. 3.7. Maximum and averaged quantitative SAR distributions within the torso induced by 10 mT and frequencies up to 100 kHz. Mean values refer to tissue volume averages.

$$\rho C \frac{\delta T}{\delta t} = \nabla \cdot (k \nabla T) + \rho Q + \rho S - \rho_b C_b \rho \omega (T - T_b) \tag{3.8}$$

where C is the specific heat capacity, k is the thermal conductivity, ω is the perfusion rate and Q is the metabolic heat generation rate. All b indices refer to blood constants. The additional source term S refers to the SAR resulting from the EM simulation. It is assumed that physiological parameters remain independent of temperature. This should be reasonable as long as temperature changes remain small. The temperature increase within 5 minutes field exposure was calculated, with the absorbed power applied by all drive field coils serving as input power. Six thermal point sensors have been positioned at various spots in the body (see Fig. 3.8), recording local temperature increase over time. While the temperature distribution within the body can be captured for certain time instances only, the thermo point sensors record up to 100,000 temperature values, depending on the time step of the simulation. No heat transfer to the surrounding was permitted to keep the energy within the body, referring to a worst case estimation. The grid was imported from the EM simulation.

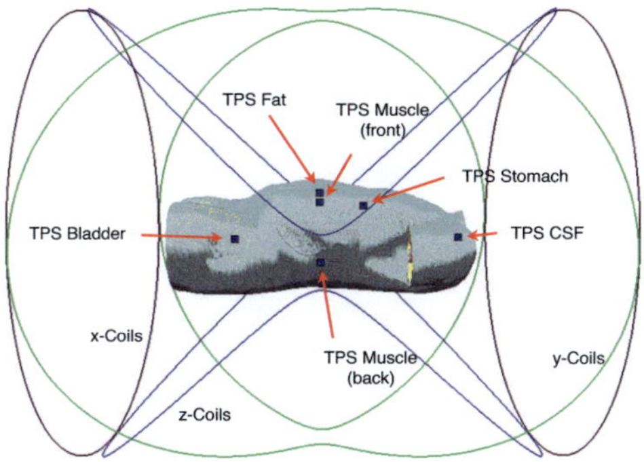

Fig. 3.8. Torso with six thermo point sensors at various positions.

Field exposure time was chosen to be 5 min, which seemed to be a reasonable duration for MPI scanning. No convective flow was integrated, i. e. the possibility of tissue cooling by moving air or liquids were neglected. *SEMCAD X* offers consideration of convection, yet the flow field is to be specified by the user. Without convection, a worst case scenario is assumed. Information about heat flow requires knowledge of tissue characteristics. Too optimistic results due to overestimating cooling can be prevented if cooling processes are neglected. Besides convective flow, *SEMCAD X* offers the possibility to specify heat transfer rates expressed by perfusion units. This cooling process comprises heat transfer to the cardiovascular system specified by perfusion rates and the specific heat capacity of blood.
In order to estimate temperature increase, comparing effects of heating and cooling terms, three simulation setups have been created:

setup A: In this setup (in the following graphs referred to as *complete model*), the extended PENNES equation is solved as stated in Eq. (3.8), considering metabolism and heat transfer to the cardiovascular system.

setup B: *(simplified model)* to keep things simple, no tissue parameters other than thermal conductivity and specific heat capacity have been considered. Neither metabolic heating nor heat transfer to the blood system was allowed. This simplifies Eq. (3.8) to

$$\rho C \frac{\delta T}{\delta t} = \nabla \cdot (k \nabla T) + \rho S. \tag{3.9}$$

setup 0: The magnetic field is switched off. Only natural thermal processes are active:

$$\rho C \frac{\delta T}{\delta t} = \nabla \cdot (k \nabla T) + \rho Q - \rho_b C_b \rho \omega (T - T_b) \tag{3.10}$$

Simulations on setup 0, i. e. with no external **B** field, result in body temperature increases by a noticeable amount. As will be seen in the following graphs, metabolic heating is in the range of 10 kHz **B** field effects, though in different organs. The maximum temperature increases after 5 minutes arise in bone-marrow, fat, kidneys and muscle. Averaged over the complete tissue volume, largest values arise in the kidneys. See Fig. 3.9 for details. According to simulation results, metabolic heating reaches a steady state with maximum and averaged temperature increase of 2.1 K and 0.08 K, respectively. Since nobody keeps getting warmer without being active in any way, one can conclude that this amount of thermal energy is usually transferred via natural thermoregulatory systems which cannot be accounted for in these simulations.

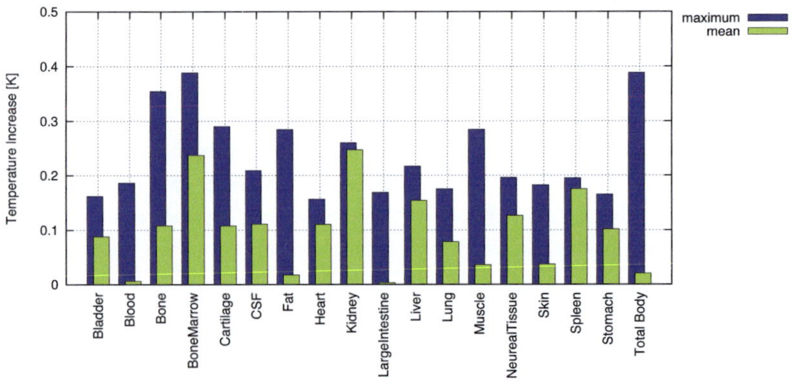

Fig. 3.9. Maximum and averaged values of temperature increase in body tissues after 300 s simulation time without external heating *simulation setup 0*. Temperature increase is caused by metabolic heating only.

At 25 kHz, the difference in temperature increase between the complete model (setup A) and the simplified model (setup B) is low: 1.53 K versus 1.46 K after 5 min field exposure. Although the generated heat can be transferred away in setup A, the absolute temperature increase is higher in the complete model due to present metabolic heating. Fig. 3.10 presents the temperature distribution as a time series of central torso slices of all three setups. One can see that thermal energy stays in the torso's periphery in the simplified model, whereas it is more evenly distributed in the complete model.

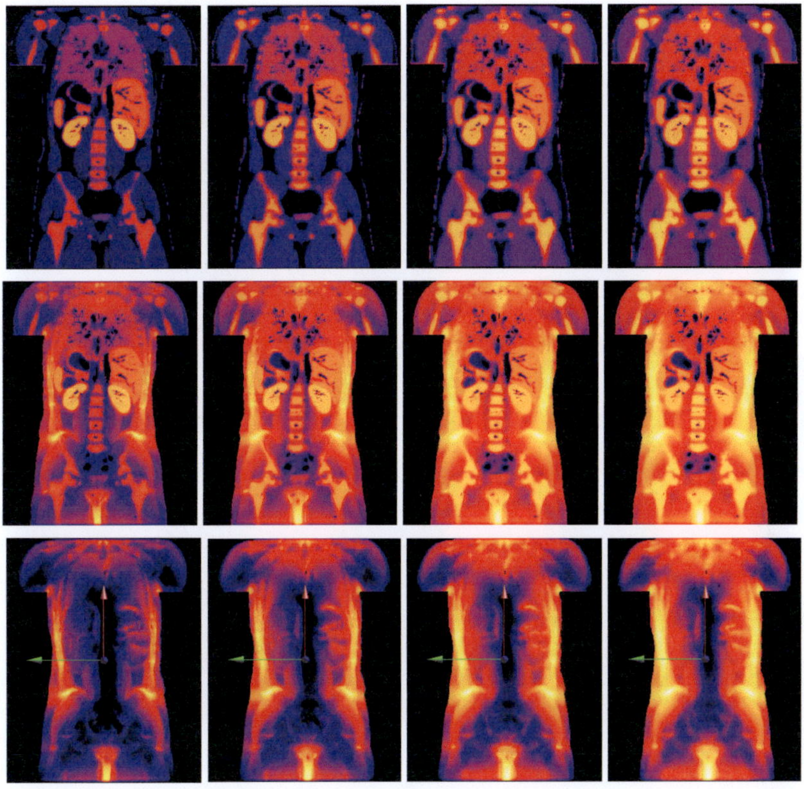

Fig. 3.10. Temperature distribution time series with **B** fields switched off (upper row, setup 0), at 25 kHz, with enabled heat transfer and metabolic heating (mid row, setup A, complete model) and with disabled heat transfer (lower row, setup B, simplified model). Scales are normalized to the maximum of 1.53 K which is reached in the complete model after 300 s. Maximum increases in the setup B model and setup 0 model are 1.48 K and 0.4 K, respectively.

Detailed information about maximum and averaged values at different frequencies after 5 min simulation time in setup A and B is provided in Fig. 3.11 and Table 3.3. As expected, temperature peaks occur in peripheral fat and muscle tissue. At lower frequencies, the complete model shows higher temperatures than the simplified setup. This is due to metabolism consideration. At higher frequencies, metabolic temperature increase is small compared to the effect of disabled heat transfer. Looking at the "total body" bar of the simplified model, one can summarize, that the temperature increase rises by a factor of 10 by approximately doubling field frequency. At 100 kHz, the maximum temperatures in both setups and tissues climb up by at least 10 K, averaged values by 2 – 5 K.

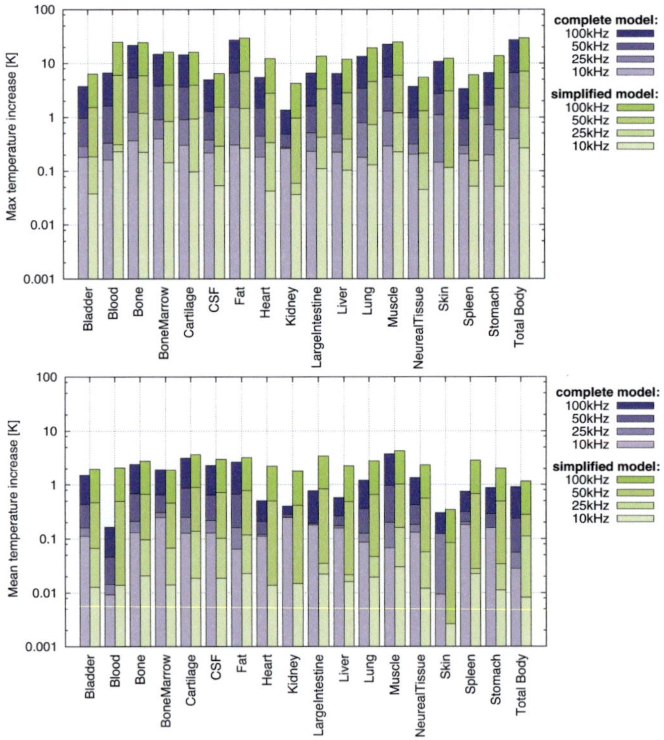

Fig. 3.11. Maximum *(top)* and averaged *(bottom)* values of temperature increase in every tissue after 300 s simulation time

		0 kHz	10 kHz		25 kHz		50 kHz		100 kHz	
		set 0	set A	set B	set A	set B	set A	set B	set A	set B
T_{max}	[K]	0.39	0.40	0.26	1.53	1.46	6.62	7.07	27.3	29.7
T_{mean}	[K]	0.02	0.03	0.01	0.06	0.11	0.24	0.28	0.91	1.16

Table 3.3. Maxima and whole body averages of induced temperature increase in setups 0, A and B.

It is well known that body tissues manage thermal energy very well. Energy applied by highly localized heat sources is transferred away fast and effectively. Maximum values calculated here might not be critical if the affected tissue volume is small enough. Fig. 3.12 presents the cumulated histograms of the temperature increases in selected body tissues at respective field frequencies. One can extract the percentage of tissue volume over temperature increase.

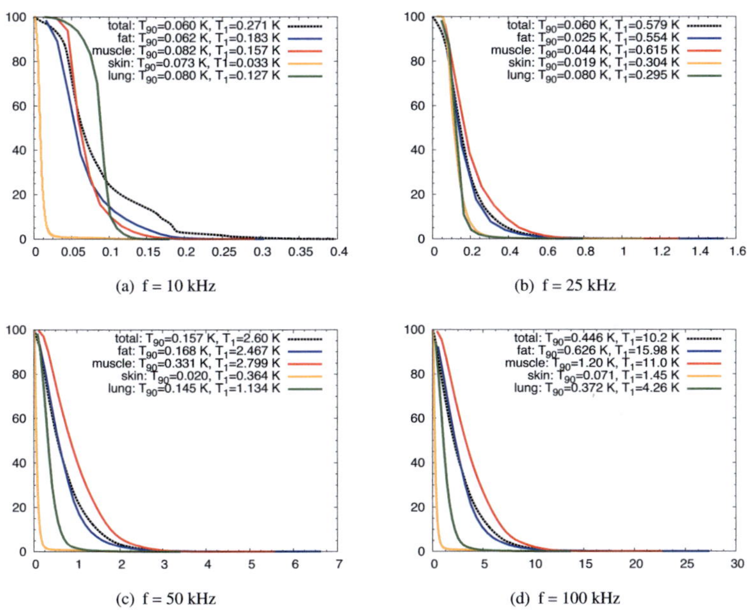

(a) f = 10 kHz

(b) f = 25 kHz

(c) f = 50 kHz

(d) f = 100 kHz

Fig. 3.12. Cumulative histograms of temperature increase in selected body tissues provided by setup A simulations. Tissue volume percentages are plotted over temperature increase. T_{90} and T_1 reflect temperature ranges occurring in 90 % and 1 % of tissue volume, respectively.

At 25 kHz, the highest calculated temperature increase appears in fat, yet only in a restricted volume of 4.5e-5 % of fat tissue, which equals 11.9 mm^3. 1 % of fat tissue is heated up by 0.55 K and more (marked as T_1 in the graph of Fig. 3.12), whereas the temperature increase in 90 % of fat tissue stays below 0.025 K. Regarding the whole torso, 1 % of torso volume is heated by 0.58 K and more, which means that 1.95 l of the torso volume get heated up by at least 0.58 K.
By positioning thermal point sensors at arbitrary spots in the body, it is possible to observe temperature increase continuously over time. Figure 3.13 illustrates temperature behavior during the course of 300 seconds simulation time.

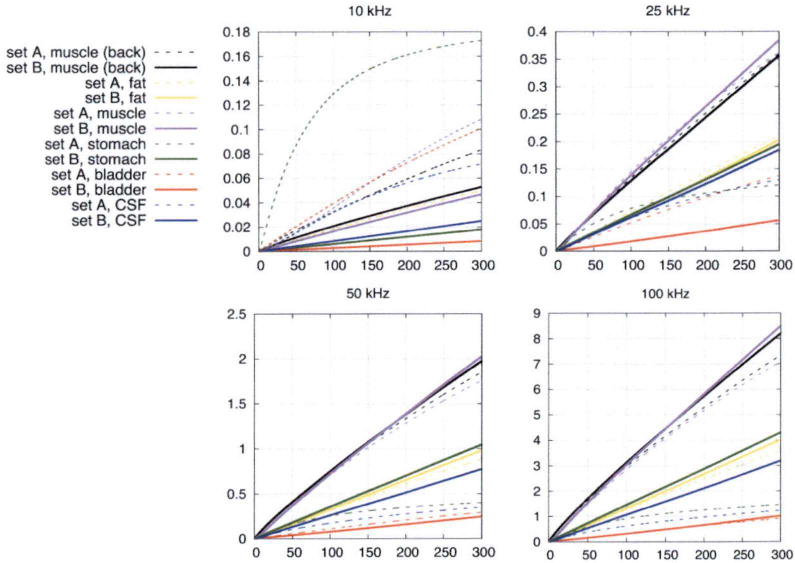

Fig. 3.13. Temperature increase (in K) over time (in s) at different simulation frequencies at various positions in the body. Dashed lines represent setup A results, solid lines refer to setup B results.

Comparing relative progresses of set A curves (complete model) with B curves (simplified model) at different frequencies, one recognizes that the angles between curves do not differ. This is not much surprising, since thermal tissue characteristics are frequency independent. What does change is the curve characteristics of set A curves (complete model). Those curves are above their corresponding sensor curve of set B at 10 kHz, but approach set B sensor curves at higher frequencies.

Depending on tissue type – and thus on perfusion rates – set A and set B curves approach each other sooner or later. Again this shows, that at higher frequencies no natural heating or cooling system has a noticable effect compared to the SAR heating.
The simplified model represents a worst case estimate. The temperatures presented will never be reached during MPI. Metabolic processes reach a steady state and can be neglected in field effect considerations. Taking perfusion rates and heat transfer into account lead to more realistic results. Still, these calculations reflect a pessimistic prediction of tissue heating, since the applied model for temperature calculations lacks in completeness of thermo-regulatory mechanisms.
Considering exposure limits, one needs to decide, whether those results satisfy requested accuracy.

3.2.3 Focus Field Induced Fields

In MPI, the drive field coils generate time-varying fields that move the field free point and excite magnetic nano-particles. In previous sections, physical effects of the kHz frequency range have been regarded. In this section, field quantities generated by the focus field will be investigated. The focus field moves the region of interest segment by segment. The field is generated by the drive field coils as well and has a frequency of $\approx$ 100 Hz. Below 1 kHz, conductivity of excitable tissue is much lower, which increases the capability of action potential generation. In contrast, heating is not an issue in that frequency range.
Fig. 3.14 presents current densities induced by 10 mT, 100 Hz. Highest current density amplitudes are induced in cerebrospinal fluid (CSF), followed by blood, muscle and intestines. Tissue volume averages show the same ranking. In crucial tissues like skeletal muscle and the heart, peaks reach values of 0.47 A/m^2 and 0.17 A/m^2, respectively, while statistical means averaged over tissue volumes are 76.5 mA/m^2 and 19.2 mA/m^2. ICNIRP guidelines [60], suggest to limit current density to 10 mA/m^2 effective value for field exposure below 1 kHz. The whole body average exceeds this value by a factor of 3.54. Considering the safety factor that has been included in the restrictions, tissue averages of J are below assume threshold amplitudes.

3.2.4 Effect of Superimposed Magnetic Fields of Different Frequencies

In acoustics, if two tones of different frequencies are played, the result is perceived as periodic variations in volume whose rate is the difference between the two fre-

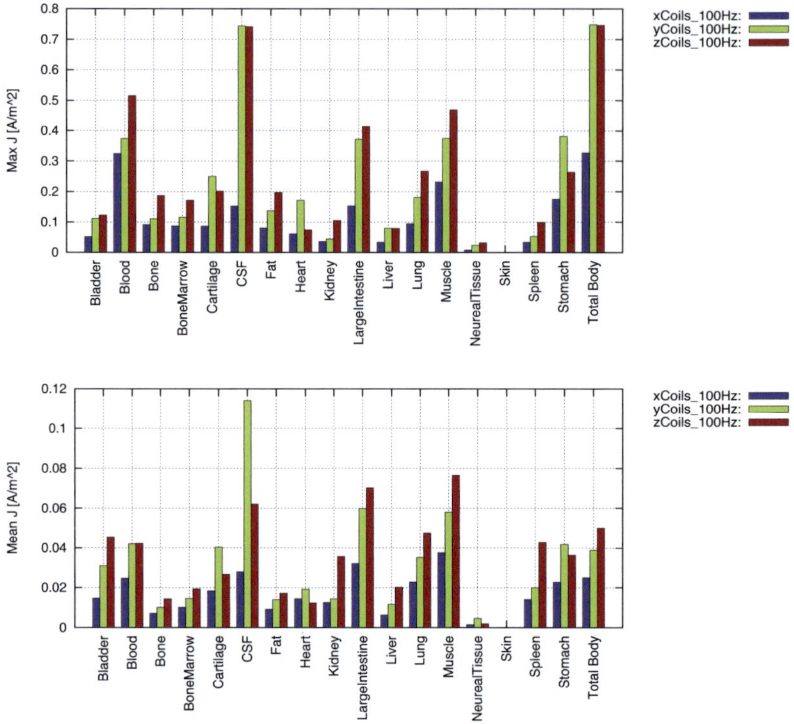

Fig. 3.14. Maximum *(upper)* and averaged *(lower)* current density distribution within the torso induced by the focus field. Mean values refer to tissue volume averages.

quencies. This phenomenon is called a *heterodyne beat* and applies accordingly to electromagnetics. The drive field x-, y-, and z-coils of the MPI system generate fields of the frequencies $f_x = 24.51$ *kHz*, $f_y = 25.25$ *kHz* and $f_z = 26.04$ *kHz*. As described in Section 2.1, this frequency shift results in a Lissajous trajectory, along which the field-free point is moved through the field of view. This frequency shift results in a heterodyne beat of both the **B** field as well as the induced **E** field and consequently the **J**, with respective beat frequency and surging amplitude. The generated **J** field will be estimated in this section.

3.2.4.1 Heterodyne Beat and Lissajous Trajectory

From theory, the superimposition of two harmonic oscillations s_1 and s_2 of frequencies f_1 and f_2 and amplitudes A_1 and A_2 results in:

$$\begin{aligned} s_{\text{ges}}(t) &= s_1(t) + s_2(t) \\ &= (A_1 - A_2)\sin(2\pi f_1 t) + A_2\left(\sin(2\pi f_1 t) + \sin(2\pi f_2 t)\right) \\ &= (A_1 - A_2)\sin(2\pi f_1 t) + 2A_2 \sin\left(2\pi \frac{f_1 + f_2}{2} t\right) \cos\left(2\pi \frac{f_2 - f_1}{2} t\right) \end{aligned} \tag{3.11}$$

Here, without loss of generality, it is assumed that $A_1 > A_2$. If $A_1 = A_2$, the first sinus term vanishes and the result is a sine wave of doubled amplitude that is enveloped by a cosine wave. Examples of beats for $A_1 = A_2$ and $A_1 > A_2$ is illustrated in Fig. 3.15.

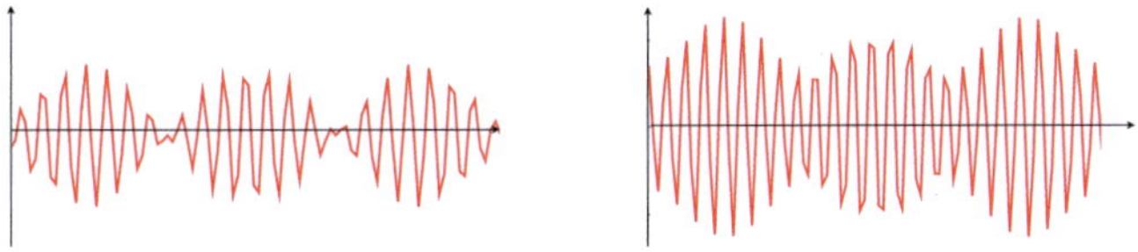

(a) $A_1 = A_2$ (b) $A_1 > A_2$

Fig. 3.15. Examples of superimposed harmonic oscillations of slightly different frequencies

If the amplitudes are vectors pointing in different directions, the resulting vector describes a LISSAJOUS figure. The appearance of the figure depends on the ratio $\Delta K = \frac{2\pi f_1}{2\pi f_2}$ and phase shift ϕ of the oscillations. For $\Delta K = 1$, the figure is an ellipse, with special cases including circles ($\phi = \pi/2$) and lines ($\phi = 0$). If ΔK is rational, with $2\pi f_1, 2\pi f_2 \in \mathbb{N}$, closed curves result. The smaller ΔK, the denser the trajectory lines. Figure 3.16 and 3.17 display three different examples of 2-dimensional LISSAJOUS figures and one 3D example, respectively.
The choice of drive field frequencies results in a beat frequency of 46.4 Hz.

3.2.4.2 Superimposing Current Density Vectors

In order to estimate the effects of superimposing three oscillating **J** fields, each generated by one coil pair, the final **J** field at every point (x, y, z) in the human

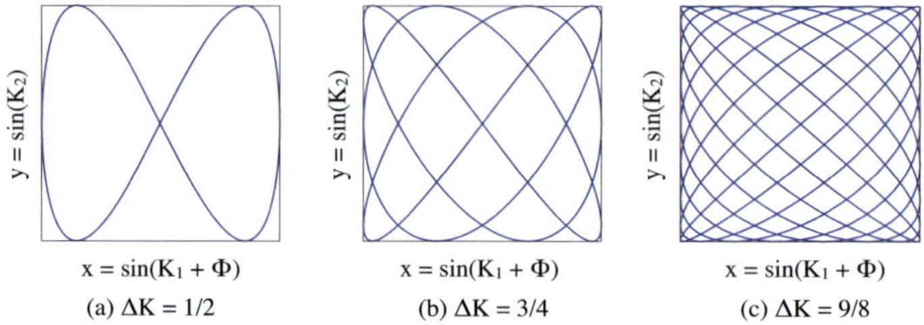

Fig. 3.16. Examples of 2D *Lissajous* trajectories

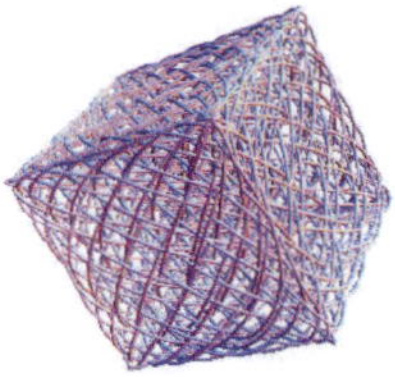

Fig. 3.17. Example of a 3D *Lissajous* trajectory

body model needs to be calculated by means of vector calculus. Every coil pair contributes to every vector component according to:

$$J_x(x,y,z,t) = J_{x1}(x,y,z)\cdot e^{i(2\pi f_x t)} + J_{x2}(x,y,z)\cdot e^{i\left(2\pi f_y t\right)} + J_{x3}(x,y,z)\cdot e^{i(2\pi f_z t)}, \quad (3.12)$$

J_y and J_z, respectively. The amplitude of the oscillating vectors, $|J(x,y,z)|$, is given as usual:

$$|J| = \sqrt{(|J_{x1}|+|J_{x2}|+|J_{x3}|)^2 + (|J_{y1}|+|J_{y2}|+|J_{y3}|)^2 + (|J_{z1}|+|J_{z2}|+|J_{z3}|)^2} \quad (3.13)$$

Results of maximum amplitudes and local averages of respective tissue are plotted in 3.18. A cumulative histogram for the excitable tissue heart and muscle is given in 3.19.

Again, no conclusion regarding excitation effects can be drawn at this point. An estimate about the effect on alternation of the transmembrane voltage, will be given in Chapter 4.

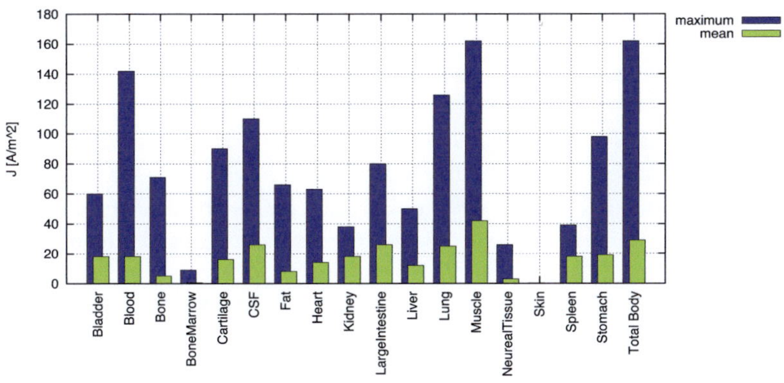

Fig. 3.18. Maximum and averaged values of induced current densities, as result of mathematical superposition of three **J** fields. The repetition time of those values is $1/46.4s = 21.5ms$.

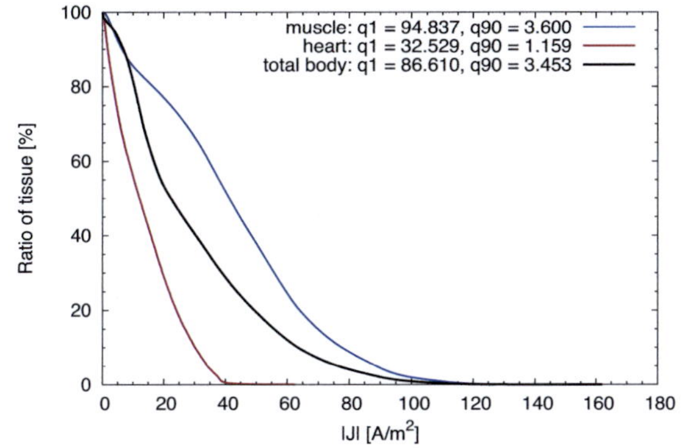

Fig. 3.19. Cumulative histogram of $|\mathbf{J}|$ in heart, muscle tissue and in the whole body

3.2.5 Drive Field Optimization

Activator of both temperature increase and current densities in the body is the electric field which is the first consequence of the applied magnetic field. Looking at the **E** field distribution, the idea arises to optimize the field distribution with regard to **E** field peak reduction. It is quite simple to generate a widely homogeneous magnetic field but the electric field distribution depends on the geometry of the field generating coils. According to Eq. (3.5), the electric field grows radially, reaching the maximum on the loop closest to the field generating loop. Thus it stands to reason to distribute the field generating current on multiple loops that produce the same target **B** field amplitudes but with smoothened **E** field distribution.

Three different geometries, displayed in Fig. 3.20, have been designed originating from the idea of distributing the applied power more evenly. To start things simple, only the $\mathbf{B_z}$-component generating coils have been considered.

version 1: 10 nested coil pairs: This arrangement consists of 10 current loop pairs with identical center and equidistant radii, starting with $r = 3$ cm of the innermost coil and ending with $r = 30$ cm of the outermost.

version 2: small body surrounding coil pairs: 24 single coils, 16 cm in diameter

version 3: small overlapping body surrounding coil pairs: 36 single coils, 16 cm in diameter.

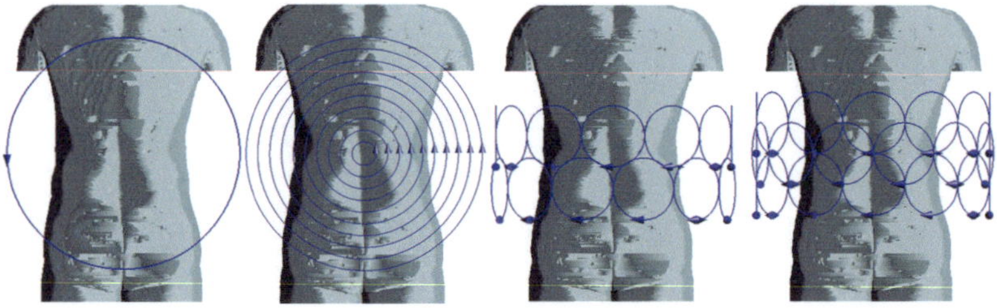

Fig. 3.20. Drive field coil configurations for field optimization. From left to right: original plain circular loop, followed by three alternative designs.

3.2.5.1 Optimization Procedure

The procedure for all suggested coil configurations was identical. First, the electric field of every singe coil pair was simulated. Since the superposition principle is valid for electric fields, and the SAR is proportional to $|\mathbf{E}|^2$, the minimizing function can be written as:

$$std \sum_{i=1}^{n} |E(x)|^2 \stackrel{!}{=} min, \tag{3.14}$$

on condition that the generated **B** field adds up to a center amplitude of 10 mT and *std* referring to the standard deviation. x represents coil currents. This constrained nonlinear optimization was solved using the sequential quadratic programming (SQP) method provided by MATLAB, giving the amplitudes of the coil currents necessary to fulfill these conditions.
The preliminary simulations were carried out on a cylindrical model, filled with saline solution, that replaced the human body model in order to speed up simulation times. From these simulations, the central **B** field vector and the **E** field vectors within a region of interest were extracted and stored. The observed region covers the area of the human body. The starting vector for the iterative optimization procedure contained the current amplitudes of every coil pair.
The optimization algorithm finds the minimum of the problem specified by:

$$\min_x f(x) \qquad \text{such that} \qquad \begin{cases} Aeq \cdot x = beq \\ lb \leq x \leq ub \end{cases} \tag{3.15}$$

where x, beq, lb, and ub are vectors, Aeq is a matrix and $f(x)$ is a function that returns a scalar. $f(x)$ can be a nonlinear function. Applied to the **E** field optimization problem with n coil pairs, Eq. (3.15) becomes:

$$\min_x \left[std \sum_{i=1}^{n} |E(x)|^2 \right] \qquad \text{such that} \qquad \begin{cases} B_n \cdot x = B_{center} \\ lb \leq x \leq ub \end{cases} \tag{3.16}$$

lb and ub define a set of lower and upper bounds on the design variables in x, so that the solution is always in the range $lb \leq x \leq ub$. Here, the boundaries were chosen such that coil currents did not fall below 0 A and not exceed 10^6 A. B_n represents the central **B** field vectors generated by the respective coil pair:

$$B_n = \begin{pmatrix} B_{x_1} & \cdots & B_{x_n} \\ B_{y_1} & \cdots & B_{y_n} \\ B_{z_1} & \cdots & B_{z_n} \end{pmatrix}$$

Thus, B_{center} is the sum of all central **B** field vectors, weighted with the target vector x. The iteration process starts with a starting vector x_0, which in this case contained the input currents of the coil pairs. Here, $x_{0_{1\cdots n}} = 1$ A. Every iteration step supplies a target vector x of length n. In the end, x contained the input factors each coil current needed to be multiplied with in order to satisfy the optimization conditions.

Version 1: 10 nested coil pairs

Fig. 3.21 displays the models for both the preliminary simulation and the final simulation for optimization validation. Optimization proceeded as described above:

- 10 simulations with the homogeneous model, running every coil pair in a separate simulation with 1 A current amplitude
- extraction of electric field, import in MATLAB
- optimization algorithm, minimizing the standard deviation of the sum of electric field vectors along the x-axis where the highest values are to be expected
- validation of optimization results by calculating the electric fields with the torso model and optimized coil currents

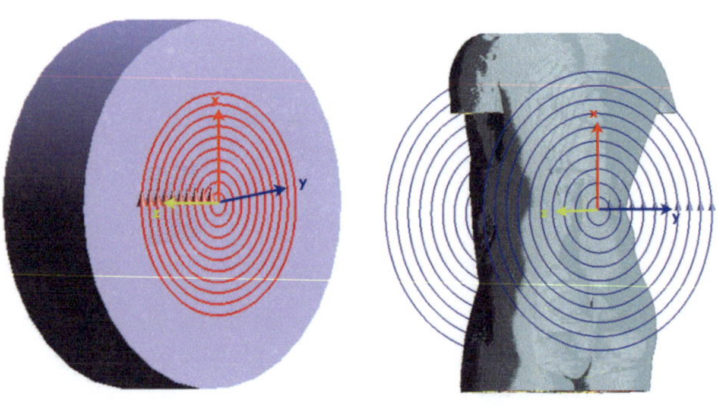

(a) (b)

Fig. 3.21. Model of coil configuration version 1: 10 nested concentric coil pairs. (a) homogeneous model filled with saline solution for the preliminary simulations, (b) target model with human torso.

Fig. 3.22(a) displays amplitudes of electric field strengths along the x-axis of the model as result of running each coil pair in a separate simulation with 1 A coil current, respectively. As expected, field strengths are highest at locations closest to the field inducing coil (coils projections are denoted by red crosses in the plot). The sum of these **E** fields is low in the center and reaches a peak within the region of interest.

After 66 iterations, the algorithm has found a minimum for the given constraints. The adapted coil currents now generate a flattened total **E** field profile. Fig. 3.22(b) shows the generated **E** field values of each coil pair and the resulting total **E** field, respectively. Due to the chosen region of interest, the outermost coils carry the largest currents. The selected slice is one close to the model surface, where the highest heating effects are to be expected. The total **E** field now is more homogeneously distributed, which will lead to reduced SAR peaks and thus reduced thermal heating.

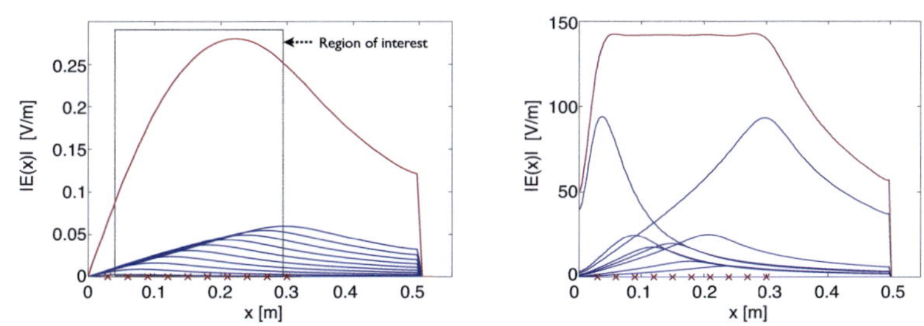

(a) results of preliminary simulations (all coils run with 1 A)

(b) results with optimized coil currents (center **B** amplitude = 10 mT)

Fig. 3.22. Electric field amplitudes in the homogeneous model along x

In order to validate these results, the torso model was exposed to the magnetic fields generated by the optimized coil currents. The result exhibits reduced power absorption all over the body, compared to field generation with one coil pair of radius $r = 30$ cm. These results can be viewed in the graphs of Fig. 3.23 and 3.24.

version 2 and 3: small body surrounding coil pairs

Motivated by the results of the nested coil geometry, alternative coil configurations had soon been designed with hope for even more success. By decreasing the coils' radii, the contribution of each coil pair to the electric field is also decreased

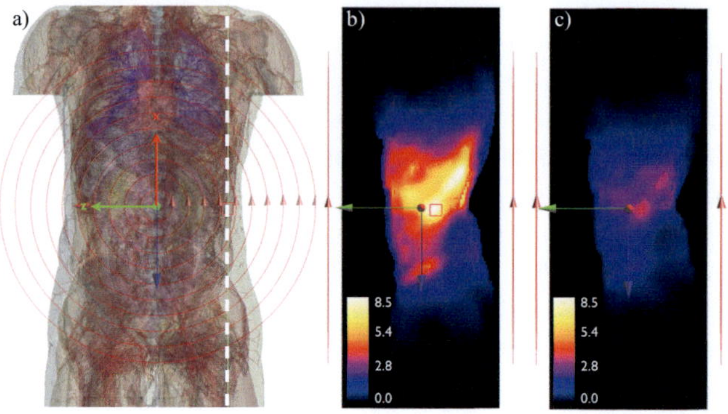

Fig. 3.23. Validation of the optimization results with human model: Generating 10 mT with one coil pair only, led to an SAR_{max} of 8.7 W/kg, shown in *(b)*, the respective slice is marked in *(a)*. Running concentric coil pairs reduces the SAR_{max} to 4.4 W/kg *(c)*.

and consequently the total electric field is more evenly distributed. Optimization procedure was straight forward like before. The preliminary simulations were carried out on a cylinder filled with saline solution surrounded by 24 field generating coils (model version 2, Fig. 3.25(a)) and 36 field generating coils (model version 3, Fig. 3.25(b)).

It turned out that the solution of these optimizing problems were not that unique compared to the nested coils configuration. The optimization algorithm found some local minima which had been taken as input for the validation with the torso model. With version 2 (12 coil pairs) the SAR could not be reduced, higher SAR levels compared to the original single-coil configuration were achieved. Utilizing 18 overlapping coil pairs, the SAR levels were only slightly reduced. Fig. 3.26 presents the best SAR values achieved. Table 3.4 summarizes the results.

For further optimization, more constraints are necessary which might be of interest for future MPI geometries. At this point, it is evident, that SAR reduction is possible by altering coil dimensions and configurations. Although SAR reduction seems to be small in version 2, it might be of great benefit depending on other circumstances.

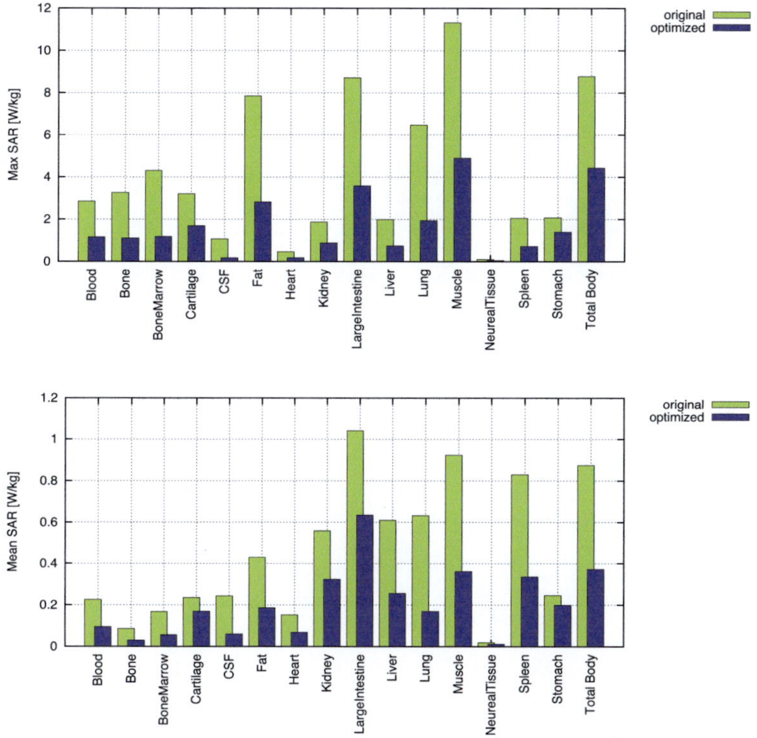

Fig. 3.24. SAR values induced by one coil pair of 30 cm radius *(original)* and by 10 concentric coil pairs with optimized input currents *(optimized)*

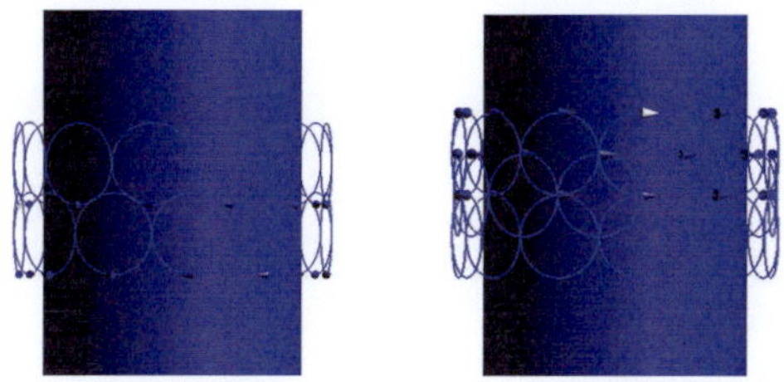

(a) vesion 2 (b) version 3

Fig. 3.25. Model of coil configuration Version 2 and 3, surrounding a water-filled cylinder

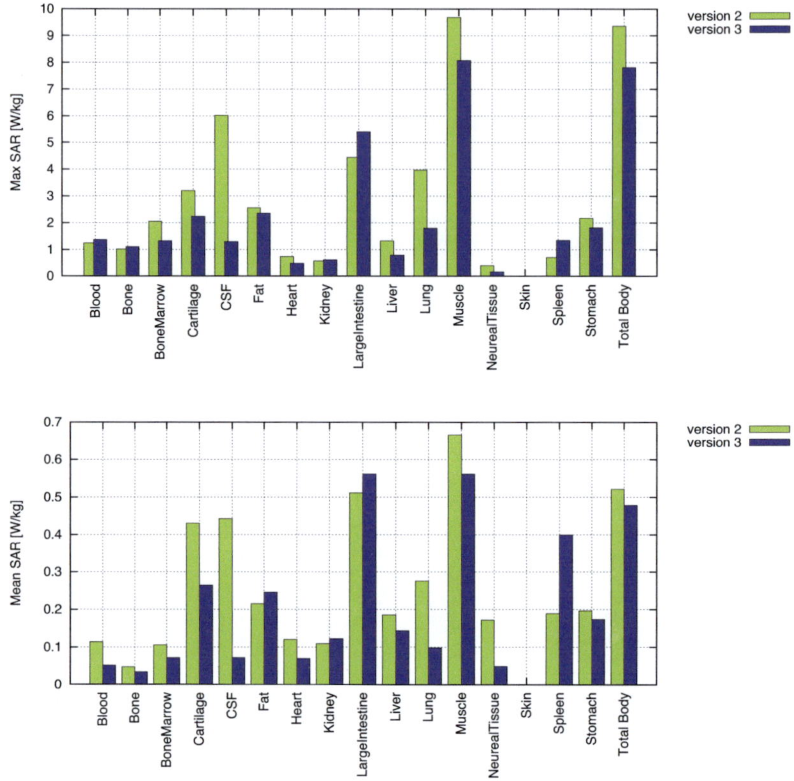

Fig. 3.26. SAR values generated by coil geometries version 2 and 3, with optimized input currents

	original single coil pair	version 1 10 nested coil pairs	version 2 12 coil pairs	version 3 18 coil pairs
Max SAR [W/kg]	8.78	4.42	9.36	7.81
Mean SAR [W/kg]	0.87	0.37	0.52	0.48

Table 3.4. Summary of the optimization results using different coil geometries

3.3 Temperature Increase Measurements in a Real MPI Scanner

The first question when thinking about temperature increase during MPI might be: Is the temperature increase measurable? In order to answer that question, a

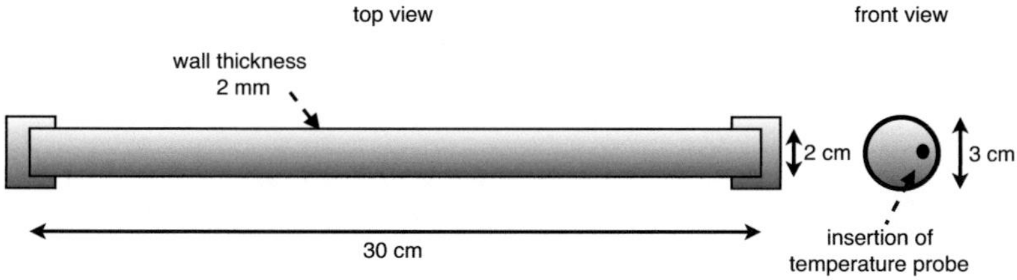

Fig. 3.27. Draft of the plexiglass phantom for temperature measurements during MPI.

phantom was built which could be filled with saline solution and – together with a fiberoptic temperature probe – inserted into the laboratory scanner at the PHILIPS research facility in Hamburg. The measurement setup, procedure and results will be described in the following sections.

3.3.1 Measurement Setup

The phantom required needed to fulfill several specifications:

- The body material had to be a poor thermal conductor.
- The phantom needed to allow complete filling with saline solution and
- positioning a temperature sensor within the phantom close to the phantom's periphery.

Fig. 3.27 presents the draft of the phantom informing about the required geometry provided by PHILIPS. The phantom finally was constructed out of plexiglass. It consisted of a cylindrical tube, closed on one end, open on the other. The lid contained a valve that allowed complete filling with saline solution and a slot for the temperature probe. Underneath the lid, a guide rail was attached, to which the temperature probe could be affixed.
The filling of the phantom was supposed to be physiological saline solution (NaCl) in order to represent human tissue. The saline concentration in physiological NaCl is 9 g/l. The temperature probe utilized was a REFLEX 4 NOTECH fiber-optic thermometer from NORTECH FIBRONIC INC, Canada. It consists of a four channel system. Each channel has a temperature resolution of 0.1 °C and an accuracy of ±1 °C plus 0.003 °C per meter of fiber. Temperature time series can be stored on a PC with a time resolution of 0.25 s.

3.3.2 Procedure and Results

For temperature measurements, the z-coils of the MPI lab scanner only had been utilized for field generation. The magnetic field was set to 20 mT, turned on and off in sequences of 5 minutes. The fiber-optic probe measured the temperature at the location of highest SAR to expect within the saline solution (see Fig. 3.28). The tube containing the probe was then inserted into the MPI scanner (Fig. 3.29).

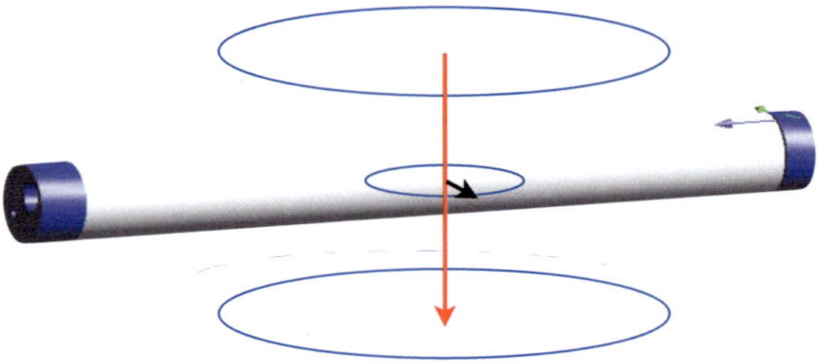

Fig. 3.28. Temperature probe position within the tube: The red arrow denotes the **B** field vector generated by the drive field coils, the blue circle within the tube marks the locations of highest induced **E** field strengths. The black arrow points at the spot, where the temperature probe was positioned during measurements.

Fig. 3.29. MPI test scanner at the PHILIPS lab in Hamburg with marked selection field coils (1), drive field coils (z-coils) (2), cooling system tubes (3) and bore hole for probe insertion (4)

During the first measurement, temperature was recorded for 15 minutes. At $t = 5$ min, the drive field coils were switched on for 6 minutes. Fig. 3.30 shows the temperature time series. The temperature almost continuously dropped by 0.25 K from start to end of the record. At first sight, it cannot be distinguished when the magnetic field had been on. The saline solution was cooled instead of heated. The cooling system of the scanner, which was set to 8 °C, had a greater influence than the warming effect of the magnetic field. Looking at the temperature differences at the moments of switching the field on and off, one recognizes that the temperature drop was larger in terms when the drive field coils were off (approx. 30 mK/min at 0 mT compared to 15 mK/min at 20 mT). This might be interpreted as "heating effect" of the absorbed power.

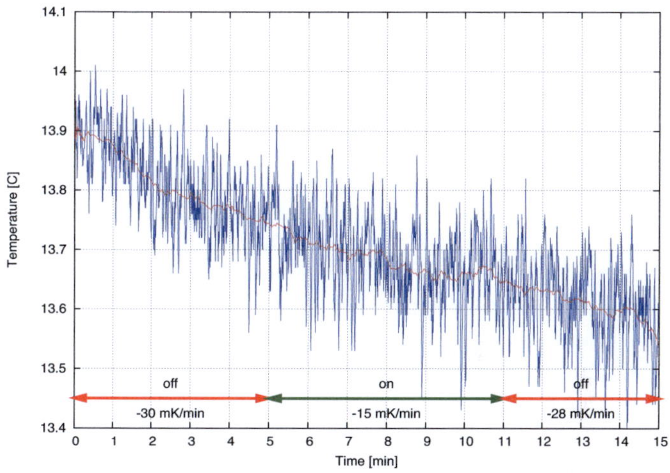

Fig. 3.30. First temperature measurement in MPI system. The red line denotes the filtered signal, obtained by applying a moving average filter to the original data. No difference is visible between coils switched on and off terms.

After several improvement steps, the actual warming effect eventually became visible. These were:

- The plexiglass tube had thermally been isolated to reduce the influence of the cooling system.
- The NaCl concentration had been increased by the factor of 20 up to 180 g/l in order to increase power absorption due to increased electrical conductivity.

- The magnetic flux density was increased up to 30 mT. However, this was not stable. After 2 minutes, magnetic flux density dropped to 20 mT.
- The periods of field exposure were extended and repeated.

From then on, the temperature drop had turned into a temperature rise during **B** field exposure terms. Fig. 3.31 shows two records. Temperature increase within field exposure varied from 60 mK (4 mK/min) up to 225 mK (15 mK/min). Obviously, the cooling system still affected the temperature inside the tube but not as strongly as before. Temperature decrease between **B** field exposure terms varied from -30 mK (-2 mK/min) to -105 mK (-7 mK/min).
These results suggest that there is no measurable temperature increase to expect during MPI sessions under normal circumstances, i. e. without increased electrical conductivity. However, within the small system, induced fields are small and so will be temperature effects. Increasing the NaCl concentration up to unnatural values finally led to a measurable signal. This was taken as proof of concept. In addition, the achieved temperature rise was used in order to validate temperature increase simulation with *SEMCAD X*, which will be presented in the following section.

3.3.3 Comparison between Measured Data and Simulation

Besides temperature measurements, simulations of the above described scenario have been carried out. All plexiglass solids have been omitted. Instead, a simple cylinder without heat transfer to bounding materials was created. The cylinder consisted of pure saline solution of increased conductivity, representing the 180 g/l-solution. Unfortunately, there was no data in literature found providing information about thermal heat capacity and thermal conductivity depending on NaCl concentration. Thus, these quantities had been kept at normal levels of $c = 4200$ J/kg/K and $k = 6.49$ W/min/K, respectively. Fig. 3.32 shows the simulation setup of the model.
The source of the heating power was the **B** field of 20 mT central amplitude generated by the drive field z-coils. For the temperature simulation, the **B** field was turned on for 15 minutes. Afterwards, temperature was recorded for another 15 minutes. Maximum temperature captured by the thermal point sensor was 0.21 K after 15 minutes heating time (see Fig. 3.33), which makes an average increase of 14 mK/min. Within the following 15 minutes, temperature dropped by 5.9 mK/min finishing at 0.122 K relative temperature increase compared to the initial temper-

ature. The simulation reconstructed the measured temperature increase and decrease perfectly fine.

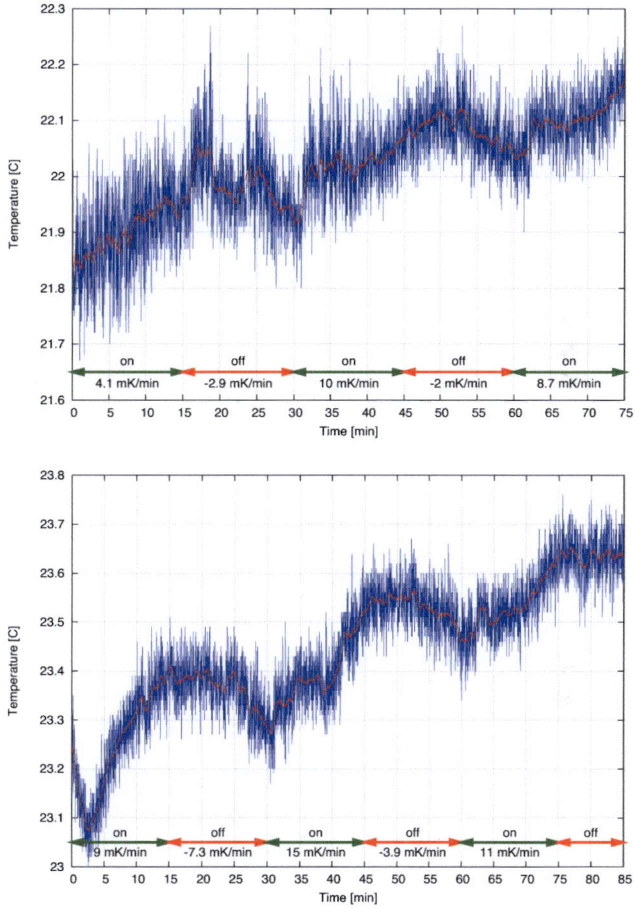

Fig. 3.31. Two records of final temperature measurements in MPI lab system

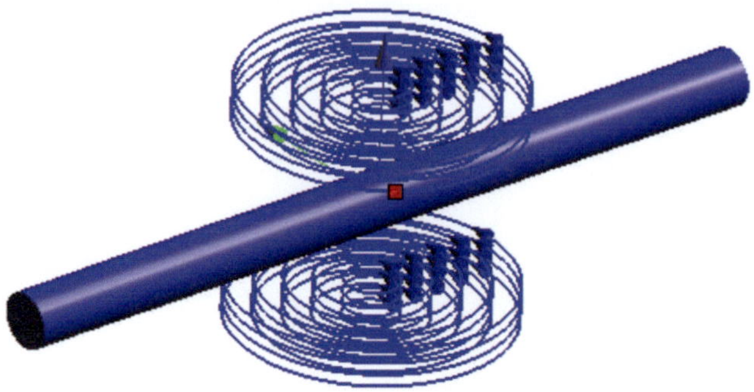

Fig. 3.32. Simulation setup of the cylindrical tube within MPI drive field coils in order to reconstruct temperature measurements. Only one solid is needed representing the saline solution. One temperature sensor is placed within the solution analog to temperature measurements.

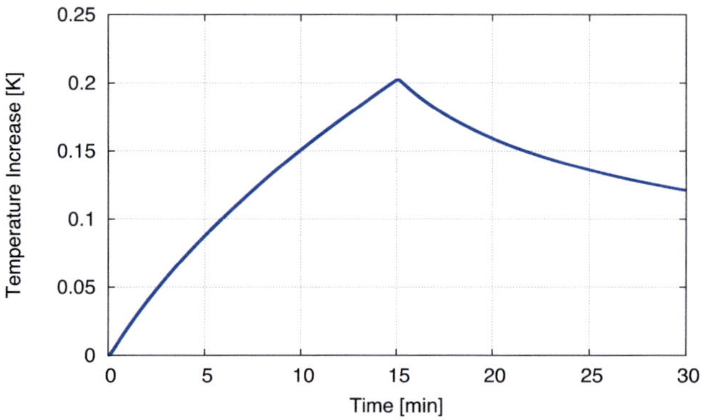

Fig. 3.33. Temperature increase captured by the thermo point sensor during simulation. Increase and decrease behavior reconstructs well measured data within the real scanner.

3.4 Results and Discussion

In MPI, crucial quantities are the induced current density, specific absorption rate and temperature increase. These quantities have been calculated with numerical methods at every point of a human model, given that the central **B** field amplitude is 10 mT.

Current density J

Due to **B** field characteristics and the given geometry of MPI drive field coils, current densities reach high values at the body's periphery. Current seeks for closed paths within conductive tissue. If paths get narrow, current flow densifies. Due to smaller possible loop radii, current densities are lower deep within the body. At 25 kHz, amplitudes of current densities reach a maximum of 63.5 A/m^2 in myocardial tissue and 168.0 A/m^2 in peripheral muscle. With increasing frequency or **B** field amplitude, current densities rise. However, these peaks are limited to very few tissue voxels. It is not clear, whether these currents are able to stimulate excitable tissues. Restrictions on current densities are defined for averaged current densities perpendicular to an area of 1 cm^2. It is well known, that currents directed along the axis of an excitable cell is more stimulative than perpendicularly oriented currents [70, 71]. At this point, limits of the utilized numerical model are reached. No assertion can be made upon the orientation between currents and muscle fibers. That's why statistical means of current densities are given here in terms of threshold considerations. Those still are 10.9 A/m^2 and 23.7 A/m^2 in heart and muscle tissue, respectively, which by far exceed current restriction of the ICNIRP (see Section 2.4.4, which is 0.36 A/m^2 for occupational exposure to electromagnetic fields of 25 kHz [60].
At 100 Hz, the frequency planned for the MPI focus field, the restrictive value of current density is 14 mA/m^2. Calculated peak and averaged values also exceed this value. Even the average over myocardial muscle is double this limit (19.2 mA/m^2). The simulation results are the basis for further investigations. Whether these calculated current densities are able to stimulate excitable tissue, i. e. peripheral nerves or even the myocardium, has to be clarified. Therefore, models of ventricular cells have been utilized to determine frequency dependent thresholds for action potential generation (see Chapter 4).

Specific absorption rate and temperature increase

The SAR is an indicator of possible temperature increase. It can be controlled by actuating the power applied to the human body. All restrictions on SAR are with regards to keeping temperature increase below 1 K. According to literature, this is given if SAR is kept below 4 W/kg [60] [62]. For electromagnetic fields below 100 kHz, no exposure limits are specified. Yet, the results presented here suggest that definitions of SAR restrictions are necessary.

Considering existing limits of field exposure, limits on local SAR are exceeded before whole body limits are exceeded. This conclusion has been drawn in studies of MRI RF coils as well [72].

At 25 kHz, maximum temperature increase values are around 1.5 K while the whole body average is below 0.01 K and thus no local burn or cell death is to be expected. This result suggests that SAR thresholds for MPI may be higher than MRI thresholds. Corresponding SAR values (25 kHz, 10 mT) are 12.1 W/kg local SAR and 0.91 W/kg whole body average.

The results show, that above 50 kHz, local temperatures might exceed 42°C after exposing the human body to a magnetic flux density of 10 mT for 5 min. In the 50 kHz-simulation, a torso volume that makes up 0.0038 % or 7.41 cm^3 exceeds 5 K temperature increase.

Above 25 kHz, the SAR has the greatest impact on body temperature compared to other temperature regulating processes like metabolism or heat transfer to the cardiovascular system. Still, both models presented here represent simplified scenarios as other thermoregulatory systems, like sweating, convective flow and others, cannot be accounted for. In order to prevent health affecting damages due to local or core body heating, it is recommended not to expose patients to 10 mT of frequencies above 50 kHz for longer than 30 seconds.

Temperature calculation validity has been proved by comparing measurements with simulated temperatures. Temperature curve characteristics show good agreement and so do values of relative temperature increase and decrease. Measured temperature increase during **B** field exposure (20 mT) is between 4 mK/min and 15 mK/min (Fig. 3.31 in sec. 3.3.2), whereas the calculated temperature rise is 14 mK/min (Fig. 3.33). The temperature drops by 2 mK/min – 7.3 mK/min in measurements compared to 5.9 mK/min calculated temperature decrease.

Alternative coil geometries are able to decrease local and global SAR values. Three options have been introduced here which give examples of different coil geometry. There is no general solution for this problem. When thinking of alternative coil configurations, other limiting factors have to be accounted for, relating to hardware issues for example. Constraining variables for **J** and SAR calculations were limited to **B** field amplitude and frequency. Appropriate **B** field generation is subject of other investigations.

Model imperfections

The human model has several drawbacks which should not be kept secret.

- The human model does not include nerves. This is probably the biggest drawback, since skeletal muscle reaction is most commonly triggered by nervous action. However, if the current density is high enough, skeletal muscle can be stimulated directly. Differences between muscular and nervous stimulation thresholds will be discussed later.
- The model does not include anisotropy, which also complicates assertions about stimulation thresholds, since stimulation effects of electric fields depend on field orientation. Again, this motivates to look into more detailed models of excitable tissue as will be described in the following chapter.
- Only the torso model was used for field effect simulations. This is due to a tradeoff that had been made during original considerations. The complete torso required a coarser grid in order not to exceed computational limits. Current density calculations showed that peak values increase, when the grid was refined. For the sake of accuracy in inner organs, especially with respect to possible myocardial stimulation, arms and legs were resigned.
- The skin is not well represented. What is assigned skin tissue is an artificial coat of one voxel layer. Hence, the skin layer in the computational grid exhibits gaps where the coat is not closed. This is most likely the reason for unrealistic low temperature results. The skin was expected to heat up more than it actually did in the simulations.

4

Behavior of the Cell Membrane of Excitable Tissue in Electric and Magnetic Fields

4.1 Introduction

In Chapter 3, current density distributions within a complete human torso data set have been presented, induced by three sets of coils which generate a magnetic flux density of 10 mT amplitude at various frequencies. The following sections evaluate the effect of current density to excitable tissue on a microscopic scale. Cell excitation is triggered by differences in charge distribution inside and outside of the cell. At low frequencies, the cell generates an action potential, if the resting potential is raised by ≈ 25 mV (see Section 2.3). Due to the capacitive characteristics of the cell membrane, the lift of the transmembrane voltage (TMV), necessary for action potential triggering, becomes more and more "demanding" with increasing frequencies. For cell excitation caused by external fields, this means that higher field amplitudes are necessary for stimulation with increasing frequency. Again, numerical calculations have been used to quantify potential distributions at the cell membranes of numerical cell representations of myocardium, skeletal muscle and nerve tissue. Both direct contact stimulation and magnetically induced field distributions have been investigated. The chapter will be closed with a comparison of numerical simulation results with a well established analytical human cell model.

4.2 Numerical Simulations upon Transmembrane Voltages

4.2.1 Setting up the Models

The cell models for the numerical simulations were created with a C++ program. Basically, they consist of coated bricks, that represent intracellular fluid within a cell membrane, embedded in extracellular fluid. 21 geometrical parameters need to be specified which are assimilated by the cell patch generator tool. The tool creates voxel-based models by successively processing through the x-, y- and z-dimension of the model, assigning material indices to the according voxels. The model specifications include:

- cell dimensions in x, y and z dimension
- cell membrane thickness
- cell-to-cell distance
- number of cells in x, y and z direction
- six material indices
- additional extracellular space bounding the cell structure
- shift between one cell layer and the following
- number of gap junctions connecting cells in x-, y- and z-direction (in case of myocardial modeling)

Geometric properties varied depending on the cell type to be modeled. Typical dimensions for representation of a myocardial cell were 50 μm $\times$ 8 μm $\times$ 8 μm. The cell membrane's thickness and intercellular gaps were set to 1 μm which exceed real cellular tissue by a factor of $\approx$ 200. This was a trade-off between modeling accuracy and numerical feasibility. Modeling a thinner membrane would increase the computational grid and with that reduce the maximum number of cells in the structure. Due to this tremendous increase of membrane thickness, the electric conductivity and relative permittivity needed to be adjusted in order to gain realistic field distributions. The adjustment procedure will be described in the following. The number of gaps varied between 0 and 30 % referring to the proportion of maximum possible gaps in the respective cell surface.
For direct contact simulations, longish cell geometries were generated. Two virtual electrodes were attached to the ends of the structure. Figure 4.1 presents examples of cell model geometries.
For inductive stimulations, patches with quadratic surface planes have been constructed (Fig. 4.2). The excitation is caused by virtual current loops which generate a magnetic flux through the cell patch and thereby an **E** field parallel to the current loops' plane.
Single motoneurons are embedded within muscular tissue. They react on external fields if they experience a change in the field pattern, either due to bending, ending or if the direction of the field changes (see Section 2.4 for details). The model shown in Fig. 4.3, despite its simplicity, is able to show this effect. The model consists of one nerve cell, coated by a cell membrane and extracellular fluid, em-

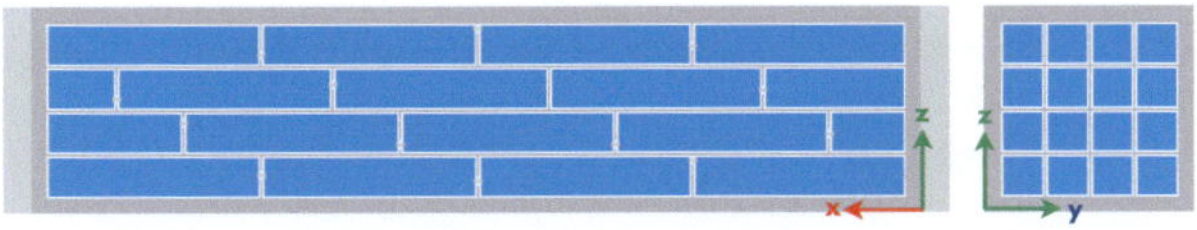

(a) Cell model for longitudinal stimulation with gap junctions in longitudinal direction

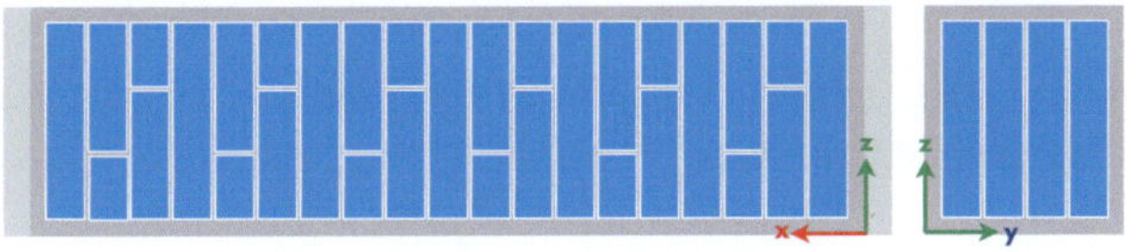

(b) Cell model for transverse stimulation with no gap junctions

Fig. 4.1. Examples of cell models for direct contact stimulations in two perspective views.

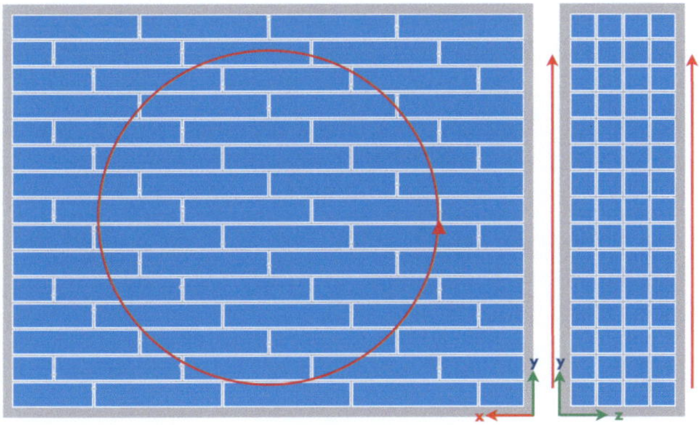

Fig. 4.2. An example of a cell patch model with current loops for field generation, top view and sagittal view

bedded in muscular tissue. Another layer of extracellular space surrounds neural tissue. Hence, current flowing from one electrode to the other first has to pass a few microns of extracellular tissue, then either enters the nerve cell through its cell membrane or enters muscle tissue, depending on the frequency of the input voltage.

This simple structures allow fast calculation of current density and potential distribution. The potential difference at a cell's membrane is decisive for action potential generation. It will be shown later, that it is possible to calculate realistic

Fig. 4.3. Simplified model of a nerve fiber within muscle tissue

field distributions despite the model's simplicity if material properties are selected carefully. Crucial quantities are

- cell and cell patch geometries
- number of gap junctions connecting cells in case of myocardial representation
- σ and ε_r of intracellular space, cell membrane, gap junctions and extracellular space

For the studies presented in the following sections, properties of the cell membrane needed to be adapted according to values of the TEN TUSSCHER ventricular model [36], which presents a highly sophisticated analytical cell model. The dielectric properties of intra- and extracellular fluids had already been investigated in previous studies at the institute [73]. In these studies, values had been initialized with equations for saline solutions [74]. Now, properties of cellular fluids had to be adjusted in order to approach GABRIEL values of myocardial, skeletal muscle and neural tissue. Within the regarded frequency range, all tissue characteristics remain constant [73].

4.2.1.1 Model Geometries

The geometries of the single cells and cell patches were chosen carefully. Myocardial cells are shorter than muscular and neural cells and are furthermore connected by gap junctions which leads to an increased conduction velocity. There are more gap junctions along the long axis of the cells than on the short axis. Gaps were spread randomly over the cell membrane area. Table 4.1 summarizes the models' characteristics. For penetration with touching electrodes, it is important that the distance between the electrodes is long enough for the current to distribute within the extra- and intracellular space. The expansion normal to the long axis can be much shorter, since the current seeks for its optimum path towards the opposite electrode. In case of magnetic sources, current loops will appear that are oriented transversely to the source magnetic field. Three layers of cells are enough, as long as there is room for circularly oriented current distribution within the transverse plains.

tissue type	cell expansions	number of cells		number of gaps
		elec. excitation	magn. excitation	
	x, y, z [μm]	x, y, z	x, y, z	x, y, z
myocardial	$52 \times 10 \times 10$	$10 \times 5 \times 5$	$25 \times 5 \times 3$	$2 \times 1 \times 1$
skeletal muscle	$105 \times 10 \times 10$	$10 \times 5 \times 5$	$25 \times 3 \times 3$	none
neural	$600 \times 10 \times 10$	1	1	none

Table 4.1. Geometries and cell configurations of the different cell types

4.2.1.2 Determination of σ and ε_r of the Cell Membrane

The membrane capacity of excitable cells is within the range of $1 - 2\ \frac{\mu F}{cm^2}$ [30,36,75], whereas the cell membrane's thickness is in the range of 5 – 8 nm [76]. The cell model of TENTUSSCHER ET AL. [36] was used here as analytical reference model. Cell membrane properties of the numerical model have been adjusted to match TENTUSSCHER model values, which are $d = 5\ nm$ cell membrane's thickness $C = 2\ \frac{\mu F}{cm^2}$ capacity, respectively (see Section 2.3.2 for details). In order to achieve appropriate conductance and capacitance despite much larger membrane dimensions in the numerical model, the following relationships have been applied: Regarding the cell membrane as a standard capacitor, its capacity can be formulated as

$$C = \varepsilon_0 \varepsilon_r \frac{A}{d}, \tag{4.1}$$

with C denoting the capacity, A the cell's surface area and d the membrane's thickness.

Solving Eq. (4.1) for ε_r assuming $d = 5\ nm$, one gets:

$$\varepsilon_r = \frac{C}{A} \cdot \frac{d}{\varepsilon_0} = 2 \frac{\mu F}{1\,cm^2} \cdot \frac{5\,nm}{8.85 \cdot 10^{-12}\,F/m} = 11.3 \tag{4.2}$$

Since

$$C_n \stackrel{!}{=} C_a \Leftrightarrow \frac{\varepsilon_{r_a}}{d_a} \stackrel{!}{=} \frac{\varepsilon_{r_n}}{d_n}, \tag{4.3}$$

with indices a referring to the analytical TEN TUSSCHER model and indices n referring to the numerical model. Solving (4.3) for ε_{r_n} yields:

$$\varepsilon_{r_n} = \frac{d_n}{d_a} \cdot \varepsilon_{r_a} = 2260. \tag{4.4}$$

Regarding the electrical conductivity of the cell membrane, the membrane conductance g_{k1} of the TENTUSSCHER-model has been reconstructed, which decisively affects the resting membrane voltage. It is given to be 5.405 nS/pF.
In general, the membrane conductivity σ can be written as:

$$\sigma = G \cdot \frac{d}{A}, \tag{4.5}$$

where G is the membrane conductance. With G_a, given by

$$G_a = C \cdot g_{k1}, \tag{4.6}$$

and C_A defined by

$$C_A = \frac{C}{A}, \tag{4.7}$$

Equation (4.5) becomes

$$\sigma_a = A \cdot C_A \cdot g_{k1} \cdot \frac{d_a}{A}. \tag{4.8}$$

Using given values, one gets:

$$\sigma = 2\,\frac{\mu F}{cm^2} \cdot 0.5.405\,\frac{nS}{pF} \cdot 5\,nm = 5.405 \cdot 10^{-13}\,\frac{S}{m} \tag{4.9}$$

Finally, requiring $G_n \overset{!}{=} G_a$:

$$\sigma_n = \sigma_a \cdot \frac{d_n}{d_a} = 5.405 \cdot 10^{-13} \frac{S}{m} \cdot \frac{1\mu m}{5nm} = 1.081 \cdot 10^{-10} \frac{S}{m} \tag{4.10}$$

4.2.1.3 Determination of Properties of Intra- and Extracellular Fluids

After having determined the electric conductivity and relative permittivity for the cell membrane, respective values need to be set for the intra- and extracellular fluids and gap junctions. The objective was to set the fluids' parameters such that the overall tissue conductivity and permittivity best match GABRIEL values for myocardial, muscle and neural material, respectively.

Both permittivity and conductivity contribute to the complex impedance of cellular tissue and can therefore be calculated using the complex current density. The complex impedance Z is represented as the equivalent series circuit of a capacitance and a resistance [26] of the cell patch:

$$Z = R + \cdot X_C = R - j \cdot \frac{1}{\omega C}, \tag{4.11}$$

with the equivalent resistance R and capacitance C.
The electric conductivity σ is defined by the real part R of the complex impedance:

$$R = \rho \cdot \frac{l}{A} = \frac{l}{\sigma \cdot A} = \frac{U}{I_{Re}} \tag{4.12}$$

Solving Eq. (4.12) for σ yields

$$\sigma = \frac{1}{\rho} = \frac{I_{Re} \cdot l}{U \cdot A}, \tag{4.13}$$

with the given quantities:

ρ : resistivity
σ : conductivity
A : cross-section of the cell patch perpendicular to the current flow
l : length of the cell patch
U : voltage (phase set to zero)
I_{Re} : current (real part)

The apparent impedance X_c and the capacitance C of the cell patch can be calculated by evaluating the imaginary part of the series impedance:

$$X_C = \frac{1}{\omega \cdot C} = \frac{U}{I_{Im}}, \tag{4.14}$$

with C defined as above:

$$C = \frac{1}{\omega \cdot X_C} = \varepsilon_r \cdot \varepsilon_0 \cdot \frac{A}{l} \tag{4.15}$$

Substitution of Eq. (4.15) into Eq. (4.14) finally yields an expression for the permittivity:

$$\varepsilon_r = \frac{I_{Im} \cdot l}{U \cdot A \cdot \omega \cdot \varepsilon_0}. \tag{4.16}$$

with I_{im} being the imaginary part of the current and ω the angular frequency $2\pi f$. Thus, for material property determination, GABRIEL values for the designated tissue types (see Table 4.2) were approached iteratively. In the first step, a frequency sweep was simulated, setting the material properties to starting values as calculated in [73]. The electrodes of either end of the cell patch were set to $0.01\,V$ and $0\,V$, respectively in case of skeletal muscle representation and 1 V and 0 V, respectively, in case of myocardial and neural representation. Simulation frequencies were set to multiples of 10, starting at 10 Hz up to 1 MHz. The obtained overall σ and ε_r were compared with GABRIEL values. From there, properties of extracellular and intracellular fluids were altered within the range of 10 % and 500 % of the original until a satisfactory similarity to the GABRIEL values was achieved.

	electric conductivity σ [S/m]			relative permittivityε_r		
f [kHz]	muscle	heart	nerve	muscle	heart	nerve
1.0e+1	0.202	0.054	0.017	2.57e+07	2.36e+07	2.01e+07
1.0e+2	0.267	0.096	0.028	9.33e+06	3.16e+06	4.66e+5
1.0e+2	0.321	0.106	0.029	4.35e+05	3.53e+05	6.99e+04
1.0e+4	0.341	0.154	0.042	2.59e+04	7.01e+04	3.56e+04
1.0e+5	0.362	0.215	0.080	8.09e+03	9.85e+03	5.13e+03
1.0e+6	0.503	0.328	0.130	1.84e+03	1.97e+03	9.26e+02

Table 4.2. Dielectric properties of excitable tissue [28]

The described scenario fulfills the quasi-static condition $\left(\frac{d}{\lambda}\right)^2 \ll 1$ (see Section 2.2.2) even for frequencies far beyond 1 MHz. In this case, *SEMCAD X* solves the electro quasi-static equation

$$\nabla \cdot \varepsilon \nabla \phi = 0. \tag{4.17}$$

The computational grid consisted of five million nodes (skeletal muscle) and 2.5 million nodes (myocardial tissue). Depending on the frequency and workstation, the simulations took 2 – 7 hours. For average σ and ε_r calculations, the resulting complex current density was extracted within the midst slice perpendicular to the potential gradient, i. e. half way from one electrode to the other (see Fig. 4.4). There, the current is supposed to have distributed evenly over the intracellular and extracellular spaces[1]. With

[1] Extracting the current density at 10 other parallel planes lead to the same averaged values of σ and ε_r,

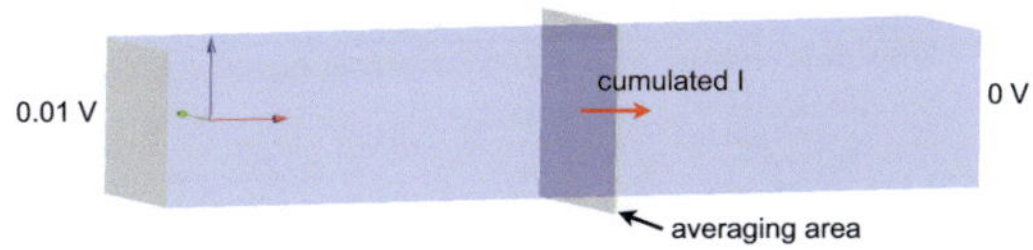

Fig. 4.4. Visualization of the slice of current density extraction for σ and ε calculations

$$I_{Re} = \int_A J_{Re}\,\mathrm{d}A \qquad \text{and} \qquad I_{Im} = \int_A J_{Im}\,\mathrm{d}A, \tag{4.18}$$

the values of I_{Re} and I_{Im} were calculated. l/A necessary for solving Equations (4.13) and (4.16) is given by the model's geometry. U is determined by the potential difference between the two electrodes; $V_{electrode_1} - V_{electrode_2}$. With that, the overall σ and ε_r could be calculated for each simulation frequency and parameter change according to Eq's. (4.13) and (4.16). Table 4.3 summarizes cellular tissue properties assigned in the numerical model as result of the adjustment simulations. Figure 4.5 shows the results of dielectric properties graphically. Myocardial values match GABRIEL values well for both σ and ε_r. Parameters for skeletal muscle are also within an acceptable range, considering that even GABRIEL values reflect only an average of measured data. Properties for neural tissue were determined accordingly, using a model that was filled with neurons only, without surrounding muscle tissue.

	extracellular fluid		intracellular fluid		cell membrane		gap junctions	
	$\sigma\,[S/m]$	ε_r	$\sigma\,[S/m]$	ε_r	$\sigma\,[S/m]$	ε_r	$\sigma\,[S/m]$	ε_r
original [73]	1.02	74.3	1.51	74.3	$5.0 \cdot 10^{-12}$	2260	1.51	74.3
heart	0.1286	74.3	0.612	74.3	1.081e-10	2260	0.612	74.3
muscle	0.643	74.3	0.612	74.3	1.081e-10	2260	–	–
nerve	0.0645	74.3	0.245	74.3	1.081e-10	2260	–	–

Table 4.3. Adapted dielectric properties of excitable tissues

Tables 4.4, 4.5 and 4.6 provide the data that were achieved after assignment of the tissue properties listed in Table 4.3. Current density values are slice maximums and means.

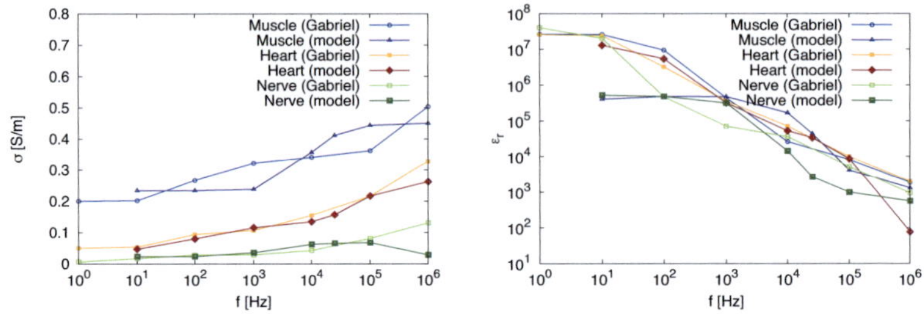

(a) Electric conductivity σ (b) Relative permittivity ε_r

Fig. 4.5. Dielectric properties of excitable tissues according to GABRIEL ET AL. and as result of parameter adaptation of the numerical models

f [Hz]	σ [S/m]	ε_r	J_{max} [A/m^2]	J_{mean} [A/m^2]	I [A]
1.0e+01	4.71e-02	1.28e+07	2.46e+02	1.13e+02	3.63e-07
1.0e+02	7.98e-02	5.37e+06	8.36e+02	1.82e+02	6.49e-07
1.0e+03	1.15e-01	3.15e+05	9.36e+02	2.26e+02	8.84e-07
1.0e+04	1.34e-01	5.19e+04	1.22e+03	2.70e+02	1.04e-06
2.5e+04	1.56e-01	3.36e+04	1.13e+03	3.22e+02	1.24e-06
1.0e+05	2.16e-01	8.57e+03	1.08e+03	4.26e+02	1.69e-06
1.0e+06	2.62e-01	7.60e+01	1.08e+03	4.95e+02	1.99e-06

Table 4.4. Resulting dielectric properties and currents within the myocardial model. Input potential drop over the model is 1 V. σ and ε are plotted in Fig. 4.5.

f [Hz]	σ [S/m]	ε_r	J_{max} [A/m^2]	J_{mean} [A/m^2]	I [A]
1.0e+01	2.34e-01	3.99e+05	6.11	2.38	8.97e-09
1.0e+02	2.34e-01	4.76e+05	6.11	2.41	8.97e-09
1.0e+03	2.38e-01	4.69e+05	6.10	2.65	9.17e-09
2.5e+04	4.11e-01	4.33e+04	6.08	4.14	1.59e-08
1.0e+04	3.57e-01	1.66e+05	5.97	3.82	1.41e-08
1.0e+05	4.43e-01	4.11e+03	6.09	4.38	1.70e-08
1.0e+06	4.50e-01	1.29e+03	5.91	4.45	1.75e-08

Table 4.5. Resulting dielectric properties and currents within the model of skeletal muscle. Input potential drop over the model is 0.01 V. σ and ε are plotted in Fig. 4.5.

f [Hz]	σ [S/m]	ε_r	J_{max} [A/m^2]	J_{mean} [A/m^2]	I [A]
1.0e+01	2.34e-02	5.19e+05	6.11e-01	2.41e-01	8.97e-10
1.0e+02	2.36e-02	4.78e+05	6.10e-01	2.66e-01	9.10e-10
1.0e+03	3.57e-02	3.13e+05	7.25e-01	4.40e-01	1.52e-09
1.0e+04	6.20e-02	1.42e+04	1.20e+0	6.13e-01	2.39e-09
2.5e+04	6.54e-02	2.65e+03	1.17e+0	6.39e-01	2.51e-09
1.0e+05	6.75e-02	9.85e+02	1.12e+0	6.54e-01	2.60e-09
1.0e+06	2.82e-02	5.56e+02	6.81e-01	4.03e-01	1.60e-09

Table 4.6. Resulting dielectric properties and currents within the neural tissue model. Input potential drop over the model is 0.01 V. σ and ε are plotted in Fig. 4.5.

4.2.2 Transmembrane Voltages Caused by Touching Electrodes

Regarding muscle and heart tissue, results of the simulations described in the previous section also serve as results for the investigation of potential distributions due to direct contact of excitable tissue with electrodes. For neural behavior observations, the initially introduced single cell model has been used for threshold determination. In order to evaluate the stimulating effect, potentials ϕ were extracted along the direction of the potential gradient from one electrode to the other. ϕ_e denotes the extracellular potential, extracted from the extracellular space neighboring the midst cell of the model. ϕ_i presents the intracellular potential of the same cell. Hence, the difference between both $\phi_i - \phi_e$ provides the transmembrane voltage V_m. Fig. 4.8(a) illustrates the procedure: with x denoting the axis directed along the potential gradient, V_m is obtained via $\phi(x, y-1) - \phi(x, y)$. It must be considered, however, that without an external voltage source, all tissues in these models are on the same potential level, in contrast to the nature of real cells. The resting potential is always to be added to the results of the following figures. Hence ϕ_e and ϕ_i and consequently V_m, are to be taken as "delta" to the resting potentials. At low frequencies, the potential drop is restricted to the extracellular space, since the membrane isolates the cell bodies. With increasing frequency, current passes the membrane and hence decreases the potential difference between inside and outside.

Figure 4.6 visualizes the effect in myocardial representation: The potential distribution along the potential gradient is shown for $f = 10$ Hz and $f = 1$ kHz. At 10 Hz (Fig. 4.6(a)), all intracellular tissue is at the same potential, due to intercellular connections via gap junctions. Almost no current flows fthrough the intracellular space. In extracellular tissue, the potential falls linearly from left electrode to right electrode potential. At 1 kHz (Fig. 4.6(b)), a potential difference becomes evident only close to the electrodes. Towards the center of the patch, the potential has distributed evenly to the intra- and extracellular space. The resulting current (Fig. 4.6(c)) now enters the intracellular space and finds a low resistant path through the gap junctions.

In the model representing skeletal muscle tissue, things are a little different At 10 Hz (Fig. 4.7(a)), every cell along the potential gradient is on a different potential level, due to the absence of intercellular connections. When the current manages to pass through the membrane, which starts at $\approx$ 1 kHz (Fig. 4.6(b)), a small current flows within one cell from one end to the other. Current patterns look the same all over the muscle patch. Most of the current is still present outside the cells.

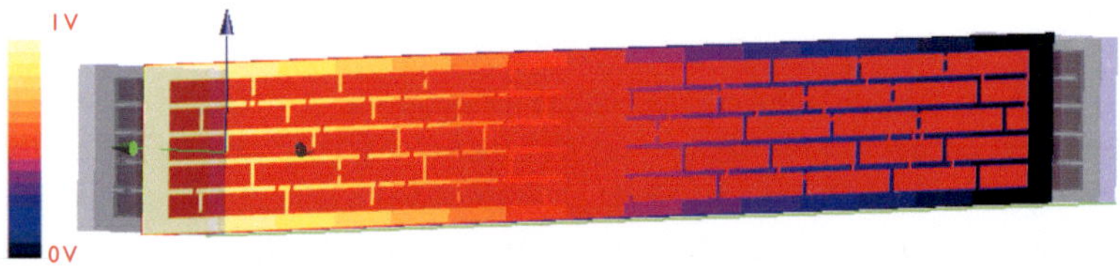

(a) Potential distribution at 10 Hz (linear scale)

(b) Potential distribution at 1 kHz (linear scale)

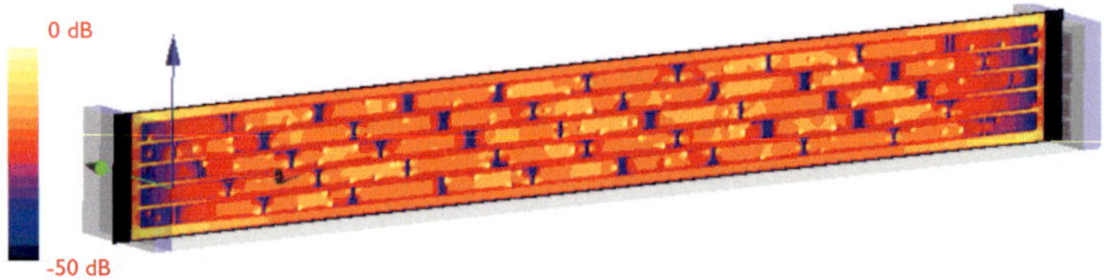

(c) Distribution of current density at 1 kHz (dB scale)

Fig. 4.6. Qualitative distribution of potential and current density within the myocardial model. The potential drops from 1.0 V to 0 from left to right. Potentials are on a linear scale, current density in dB.

(a) Potential distribution (linear scale)

(b) Current density distribution (dB scale)

Fig. 4.7. Qualitative distribution of potential and current density within the skeletal muscle model at 1 kHz. The potential drops from 0.01 V to 0 from left to right.

It is well known that stimulation thresholds are lower if the electric field is oriented along the cell's long axis compared to perpendicularly oriented fields [38, 39, 41]. In order to reconstruct this effect by means of numerical simulations, both the cells within the myocardium model and the skeletal muscle model have been rotated such that the potential gradient is normal to the long axis of the cells (see Fig. 4.1(b)). For those tissue representations, ϕ_e and ϕ_i have been extracted at points of normal cell surfaces. This is visualized in Fig. 4.8(b). V_m is the result of $\phi(x+1,y)-\phi(x,y)$ (proximal end) and $\phi(x-1,y)-\phi(x,y)$ (distal end).
In the following, the potentials ϕ_e and ϕ_i and the resulting V_m are presented in separate plots for each simulation frequency and cell patch model.

Myocardial Tissue

The first series (Fig. 4.9) presents the potential distributions of the myocardial cell patch model along the distance from one electrode ($U = 1\,V$) to the other (ground). The potential gradient is directed along the long axis of the cells. The stimulation frequency and resulting averaged current density is given in the curves' legends. ϕ_i is interrupted by intercellular gaps, whereas ϕ_e is interrupted by gap junctions which happened to show up on the selected extraction line.
Current always takes the way of lowest resistance. At very low frequencies, the current flows through the extracellular fluid only. This leads to high potential dif-

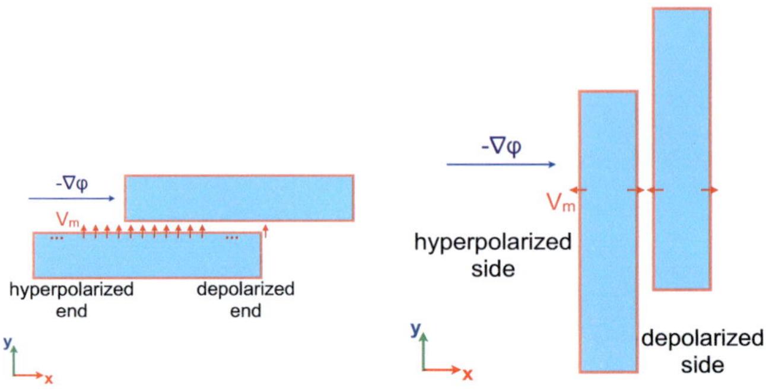

(a) (b)

Fig. 4.8. Locations of potential extraction for parallel (a) and transverse (b) stimulation

ferences at the cell membrane in planes very close to either electrode, which would force the Na^+ channels within the cell membrane to open and thereby trigger an action potential. Within the myocardium, electric signals are conducted from one cell to the next along the conduction system of the heart. This means that it is sufficient to stimulate few cells to trigger a depolarization wave. At frequencies below 1 kHz, all cells in the first half of the patch is hyperpolarized all along the length of each cell. When the extracellular potential crosses the intracellular potential level, the interior side of the cell membrane is on a higher potential level, V_m switches its sign.

With increasing frequency, ϕ_e and ϕ_i more and more approach each other. This decreases the V_m and disables the effect of cell excitation. J_{mean} increases, due to increasing overall conductivity of the tissue. Note the different scaling in the V_m plots. V_m peaks at either ends of the model disappear. Instead, all cells experience a hyperpolarization on the proximal end (close to the exciting electrode) and a depolarization on the distal end. This implies that all cells would excite simultaneously in case of super-threshold stimulation, instead of subsequently as in low frequency stimulation.

The second and third series (Figs. 4.10 and 4.11) present simulation results of the myocardium model, in which the potential gradient hits the long side of the cells. Fig. 4.10 shows ϕ_e, ϕ_i and V_m of the distal side of the cell. In this simulation, the left and right electrode were set to $+0.5$ V and -0.5 V, respectively. The intracellular

potentials differ from one cell to the following already al $f = 10$ Hz. This is due to fewer intercellular connections along the long side of the cells.
At higher frequencies, these faces of the membrane all get depolarized, due to the potential drop outside the cell. In case of sufficient TMV increase, exceeding the threshold potential, an action potential would be triggered. 48 points denote potential extraction sites of all 48 cells in x direction. Compared to longitudinal stimulation, generated TMVs are lower.
Figure 4.11 shows extractions of potentials at the proximal ends of the cells. At frequencies below 1 kHz, this side is also depolarized. Above 1 kHz, V_m of all cells is distracted to negative levels.

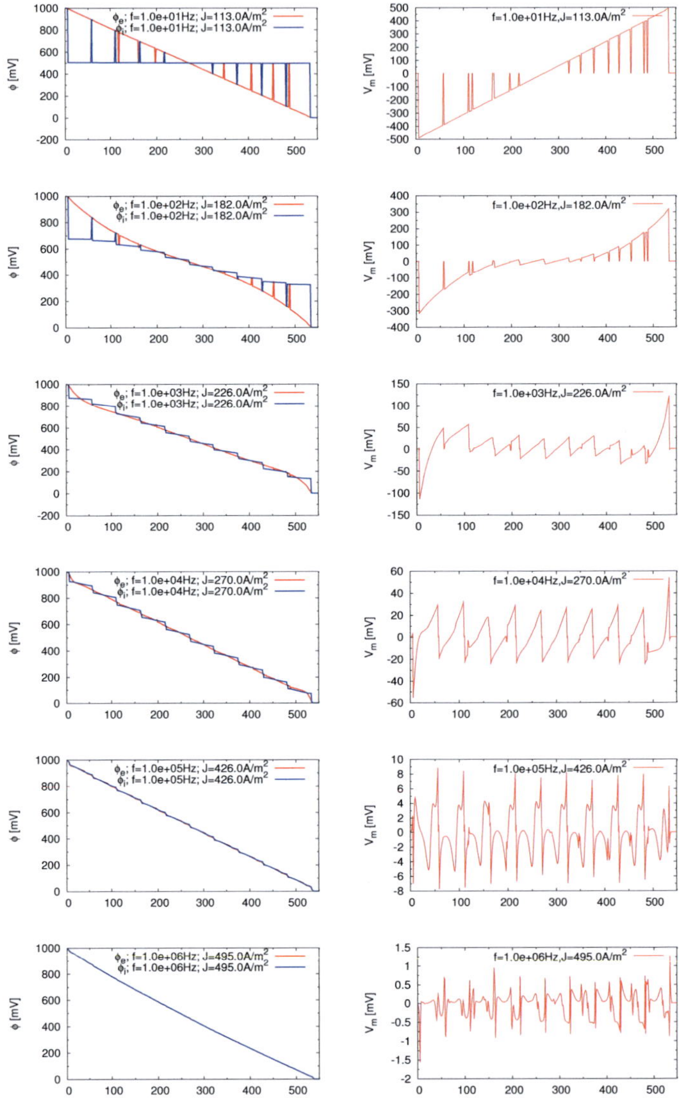

Fig. 4.9. ϕ_i and ϕ_e (left) and the corresponding TMV (right) along the x-axis (in [μm]) in the myocardium model with the cells' long axes parallel to the field gradient at frequencies from 10 Hz to 1 MHz.

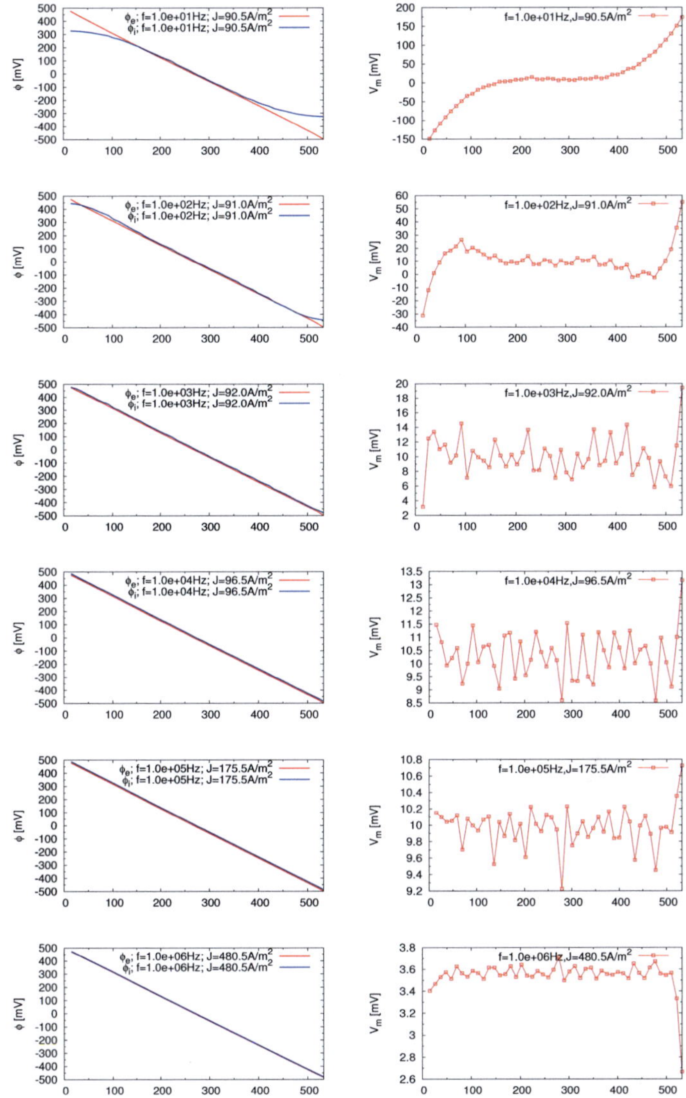

Fig. 4.10. ϕ_i and ϕ_e and the corresponding TMV along x (in [μm]) extracted at the distal ends of the myocardium model, field normal to the cells' long axis at $f = 10$ Hz up to $f = 1$MHz.

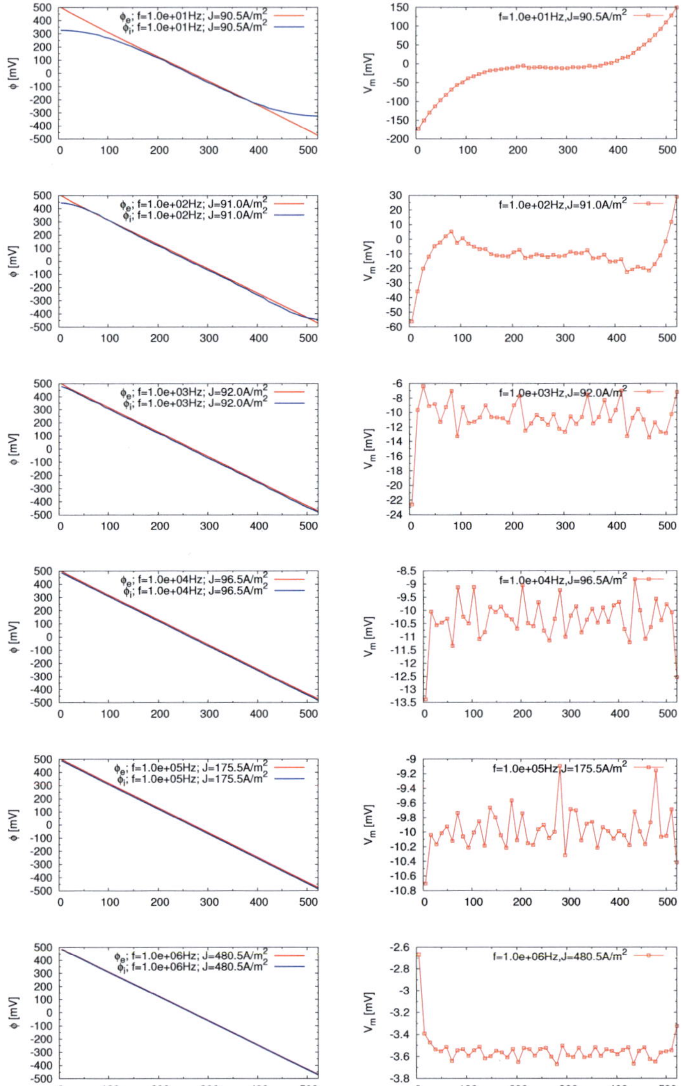

Fig. 4.11. ϕ_i and ϕ_e and the corresponding TMV along x (in [μm]) extracted at the proximal ends of the myocardium model, field normal to the cells' long axis at $f = 10$ Hz up to $f = 1$MHz.

Muscle Tissue

Looking at extracted potentials from the muscle representation, things are a little different. Cells are not connected with their neighbors. Hence every cell interior is on a different potential level, resulting in a staircase pattern of ϕ_i in case of longitudinal stimulation. This leads to the saw tooth pattern of V_m already at low frequencies and hence suggests simultaneous excitation if the increase in V_m is sufficiently high. With increasing frequency, a potential drop and thus a current flow within the cells takes place. ϕ_i and ϕ_e approach each other until they come to lie on one another. Only the cells' endings experience a potential difference to the outside. This is where the current has to pass the cell membrane. This effect again decreases with increasing frequency, since the current eventually is able to pass the cell membrane. Figure 4.13 presents the results, scaled by a factor of 100 in order to compare with myocardium model results.
Next, Fig's. 4.14 and 4.15 present the results of transverse excitation. 100 cells in the direction of the field gradient fit into the model in this configuration, delivering many points of potential extractions. Both proximal and distal faces of the cells are at a lower potential level than the extracellular space. The intracellular potential also depends on the size of the cell. Due to the shift in the cell alignment, cells differ in their shapes periodically. Cells with biggest surface area show lower intracellular potentials. This can be seen more clearly in Fig. 4.12. This results in an alternating V_m, throughout all frequencies. In general, potential differences are much lower than in the simulations of longitudinal stimulation. This goes along with findings of earlier studies.

Fig. 4.12. Potential distribution in the skeletal muscle patch model, with cells oriented such that the potential gradients hits the cells transversely. Intracellular potentials not only depend on the location relative to the voltage source, but also on the size of the affected surface area.

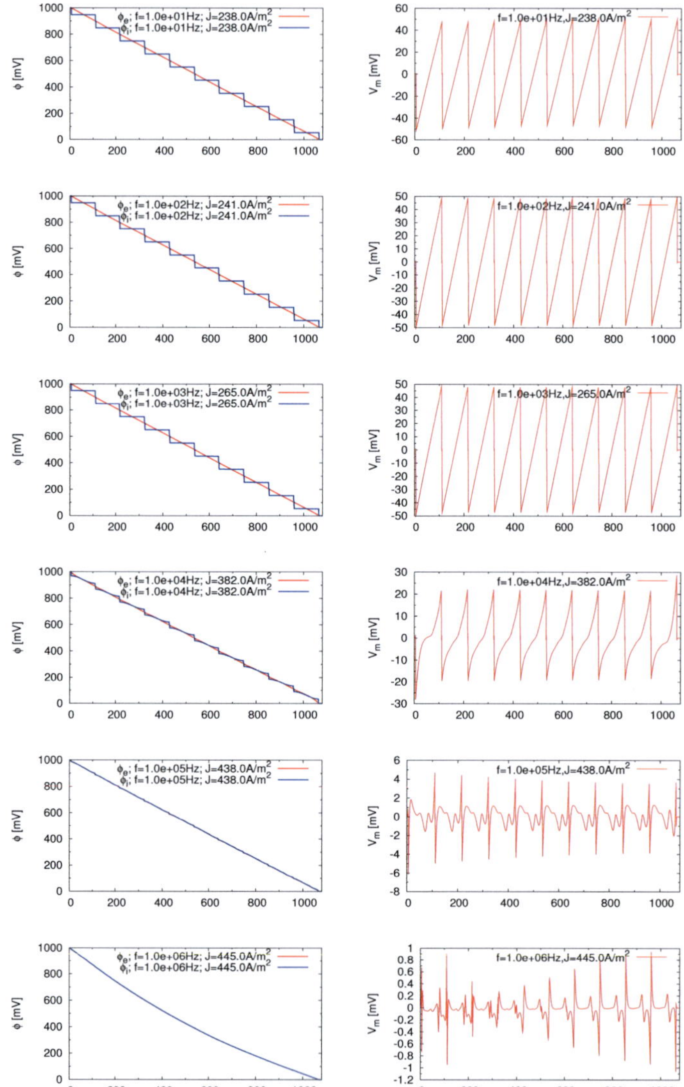

Fig. 4.13. ϕ_i and ϕ_e and the corresponding TMV along x (in [μm]) extracted along the long side of the skeletal muscle model cells, field parallel to the cells' long axis at $f = 10$ Hz up to $f = 1$MHz.

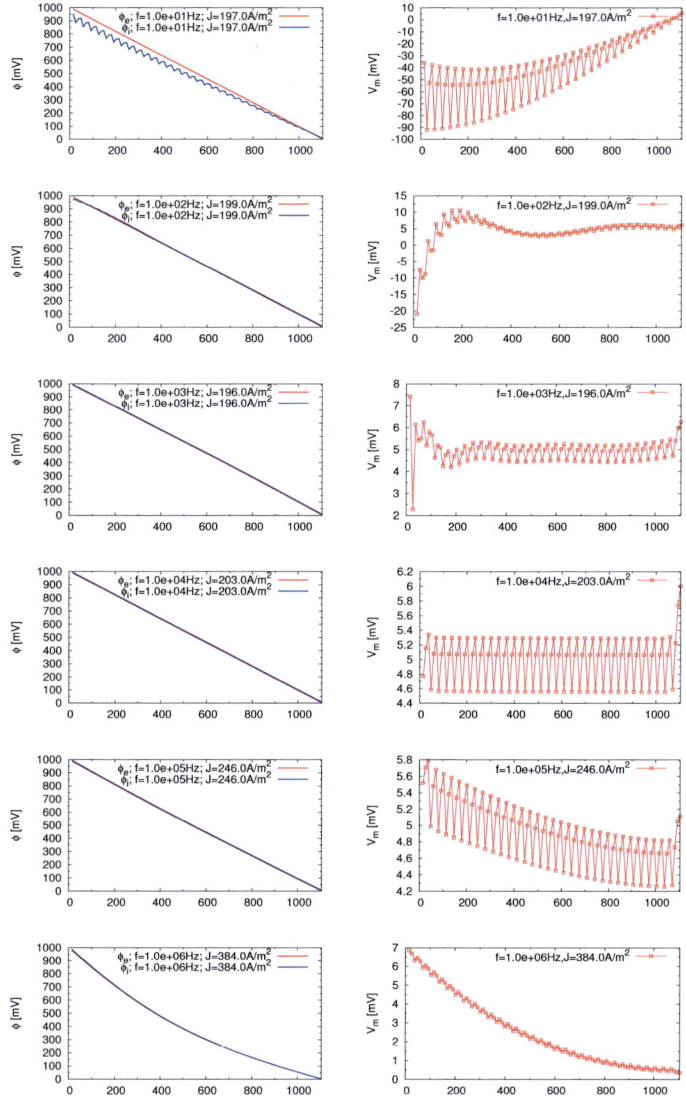

Fig. 4.14. ϕ_i and ϕ_e and the corresponding TMV extracted along x (in [μm]) at the distal ends of the skeletal muscle model cells, field normal to the cells' long axisat $f = 10$ Hz up to $f = 1$MHz.

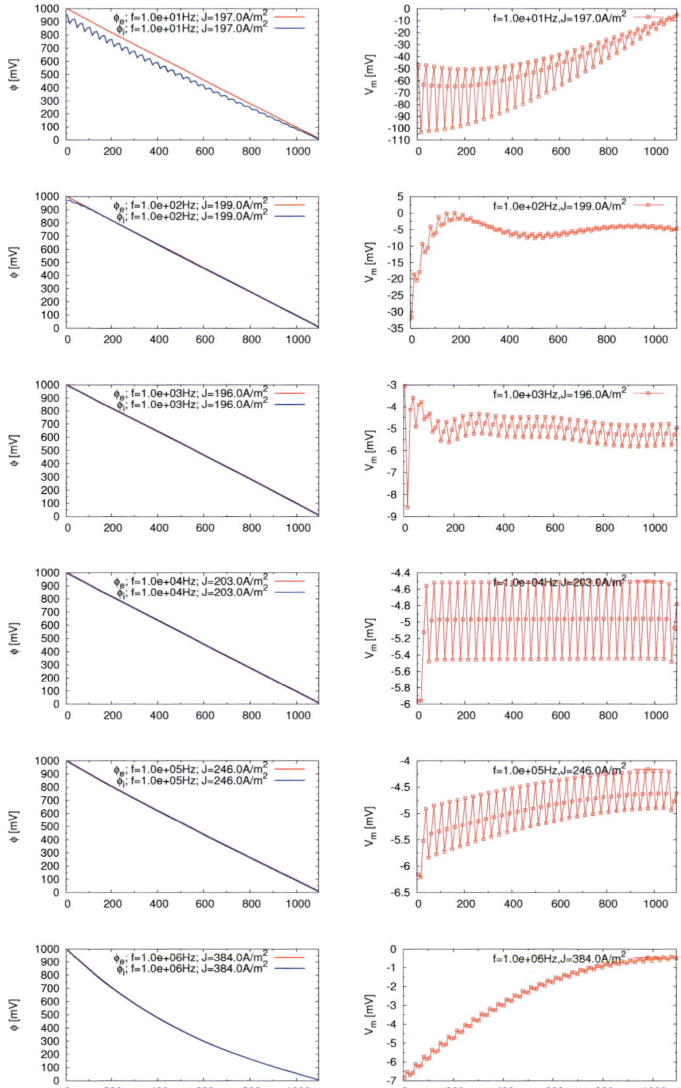

Fig. 4.15. ϕ_i and ϕ_e and the corresponding TMV extracted along x (in [μm]) at the proximal ends of the skeletal muscle model cells, field normal to the cells' long axisat $f = 10$ Hz up to $f = 1$MHz.

Single Nerve Model Embedded in Homogenous Muscle Tissue

The characteristics of the potential distribution within the nerve, displayed in Fig. 4.16, are similar to those of single muscle cells: At very low frequencies, no field is evident within the cell and no current enters the cell. Instead, current flows in muscle tissue along the nerve's boundaries, causing high potential differences at the cell membrane. With increasing frequency and current flow within the cell, a depolarization takes place only at the cell termination. Above 1 kHz, in addition to depolarization and hyperpolarization at the nerve's ends, the opposite effects appear close to the ends. The areas of de- and hyperpolarization get smaller with increasing frequency. The area of hyperpolarization next to the depolarization is called *virtual anode*. The opposite effect consequently is called *virtual cathode*.

Determination of Threshold Current Densities

For action potential generation, it is sufficient if the threshold potential (TMV_{thresh}) is exceeded at small portions of the cell's surface. $TMV_{thresh} \approx 25$ mV above the cell's resting potential. The crucial point is the current density in the tissue induced by the magnetic field, which is responsible for the TMV alternation. Therefore, the volume average current densities have been extracted from the simulations, in correspondence with obtained q10 and q1 values of TMVs at the cell membranes. Due to the linear relationship, threshold current densities can be calculated by scaling up:

$$J_{thresh,q10} = \frac{TMV_{q10}}{TMV_{thresh}} \cdot J_{avg} \tag{4.19}$$

$J_{thresh,q1}$ was calculated correspondingly. Tables 4.7, 4.8 and 4.9 provide TMV quantiles and respective current densities. It turned out that the increase in current density does not differ significantly, forcing 10 % of the TMV values to rise above threshold (q10 value of TMV) or only 1 % (q1 value). Due to the characteristic conduction system of myocardial tissue, stimulation thresholds are much lower compared to the values calculated for skeletal muscle tissue.
Respective threshold current density amplitudes are also given in Fig. 4.17 together with thresholds for transverse excitation. The trend of rising threshold currents with increasing stimulation frequency is definite, except for few outliers. Threshold current density for nerve stimulation above 100 kHz seems to decrease. This originates from the parameter adaption for dielectric properties (see Fig. 4.5). The overall tissue conductivity decreased for frequency rising above 100 kHz instead of further decreasing, as it is supposed to.

f [Hz]	J_{avg} [A/m^2]	TMV_{q10} [V]	$J_{thresh,q10}$ [A/m^2]	TMV_{q1} [V]	$J_{thresh,q1}$ [A/m^2]
1.0e+01	113.0	4.39e-01	6.44e+00	4.83e-01	5.85e+00
1.0e+02	182.0	2.80e-01	1.63e+01	3.10e-01	1.47e+01
1.0e+03	226.0	1.04e-01	5.46e+01	1.09e-01	5.17e+01
1.0e+04	270.0	3.85e-02	1.76e+02	3.85e-02	1.76e+02
1.0e+05	426.0	7.79e-03	1.37e+03	8.19e-03	1.30e+03
1.0e+06	495.0	7.36e-04	1.68e+04	7.36e-04	1.68e+04

Table 4.7. J_{avg}, 10 % quantile and 1 % quantile of the transmembrane voltages (TMV_{q10}, TMV_{q1}) calculated for longitudinal stimulation of myocardial tissue. $J_{thresh,q10}$ and $J_{thresh,q1}$ refer to the stimulation threshold current density, assuming that exceeding the threshold potential in 10 % and 1 % of the observed patch, respectively, is sufficient.

f [Hz]	J_{avg} [A/m^2]	TMV_{q10} [V]	$J_{thresh,q10}$ [A/m^2]	TMV_{q1} [V]	$J_{thresh,q1}$ [A/m^2]
1.0e+01	2.38	4.52e-04	1.32e+02	4.96e-04	1.20e+02
1.0e+02	2.41	4.47e-04	1.35e+02	4.87e-04	1.24e+02
1.0e+03	2.65	4.41e-04	1.50e+02	4.81e-04	1.38e+02
1.0e+04	3.82	2.36e-04	4.05e+02	2.52e-04	3.79e+02
1.0e+05	4.38	3.88e-05	2.82e+03	4.22e-05	2.60e+03
1.0e+06	4.45	7.82e-06	1.42e+04	8.79e-06	1.27e+04

Table 4.8. Calculated currents and TMV quantiles for longitudinal skeletal muscle stimulation

f [Hz]	J_{avg} [A/m^2]	TMV_{q10} [V]	$J_{thresh,q10}$ [A/m^2]	TMV_{q1} [V]	$J_{thresh,q1}$ [A/m^2]
1.0e+01	24.1	2.57e-01	2.35e+00	2.83e-01	2.13e+00
1.0e+02	26.6	8.24e-02	8.07e+00	9.36e-02	7.10e+00
1.0e+03	44.01	2.23e-02	4.92e+01	2.40e-02	4.58e+01
1.0e+04	61.3	5.63e-03	2.72e+02	5.63e-03	2.72e+02
1.0e+05	65.4	1.72e-03	9.48e+02	1.72e-03	9.48e+02
1.0e+06	40.3	1.77e-03	5.68e+02	1.77e-03	5.68e+02

Table 4.9. Calculated currents and TMV quantiles for longitudinal single nerve stimulation

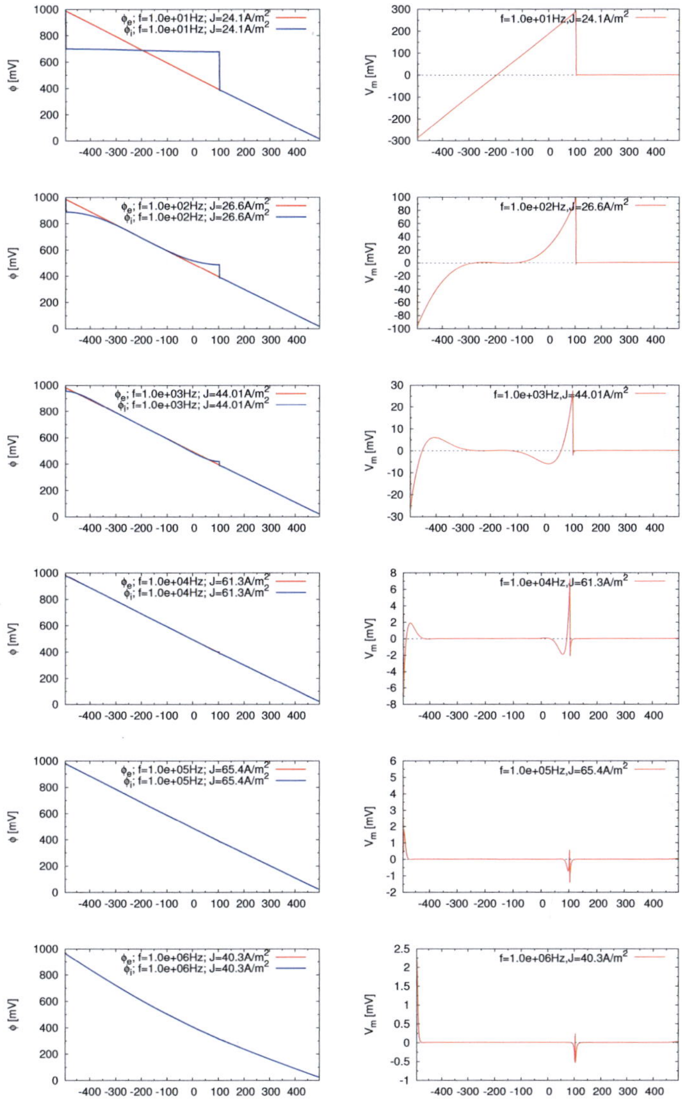

Fig. 4.16. ϕ_i and ϕ_e and the corresponding TMV extracted along the long side of the nerve cell, field parallel to the cell's long axis at $f = 10$ Hz up to $f = 1$MHz. The nerve cell terminates at $x = 100\,\mu$m.

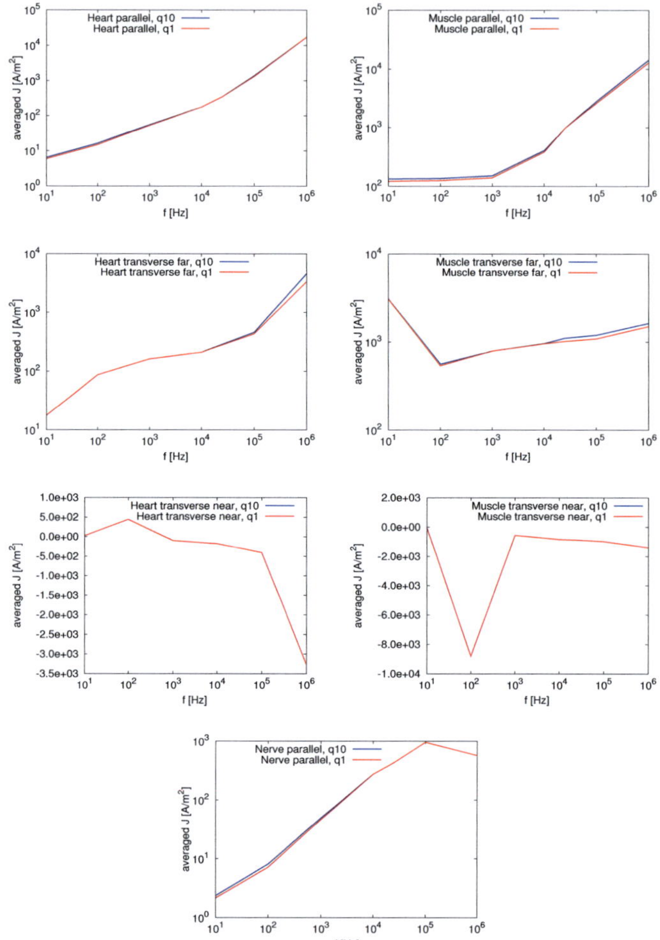

Fig. 4.17. Current density thresholds scaled to a level such that 10 % and 1 % (q10, q1) of the extracted TMV exceed the threshold for action potential generation. "parallel", "near" and "far" refer to stimulation location. "Parallel" refers to longitudinal stimulation, "near" and "far" to transverse stimulation with potentials extracted proximal and distal to the source electrode. Note that J is displayed in logarithmic scale only where appropriate.

4.2.3 Transmembrane Voltages Caused by Magnetic Fields

When stimulating excitable tissue by closed current loops, the world is a little different compared to direct contact stimulation. In direct contact stimulation, the current distributes according to the resistivity along its path, following a potential gradient. A current needs to enter the cell via the cell membrane. Magnetically induced currents within the intracellular space do not necessarily need to have past the membrane.
Due to FARADAY's law, the induced **E** field depends on the amplitude, orientation and frequency of the stimulating **B** field. Since the **E** field is rotating, no electric potential field ϕ is defined. However, the voltage across the thin cell membrane can be calculated by taking a line integral of **E** across the cell membrane. Hence, the TMV is

$$V_m = \int \mathbf{E} d\mathbf{s}, \tag{4.20}$$

with **s** being the path through the cell membrane. Regarding cell patches that are penetrated by an oscillating **B** field, the induced electric field component oriented across the cell membrane integrated over the membrane's thickness represents the TMV, that is critical for cell excitation.
The scheme of the model for induced currents simulations is that of Fig. 4.2. The coils' radius and distance is 70 μm and 60 μm, respectively. Running the coils with 1 A each results in a central B amplitude of 16 mT. In the following, resulting TMVs are presented for stimulation of myocardial and muscle tissue at 10 kHz.

Myocardial Tissue

Figure 4.18 shows three sets of plots: Figures 4.18(a) and 4.18(b) present induced E_x and E_y, respectively, by means of colored slices through the mid intersection of the model. Red denotes zero electric field, yellow and white positive values, blue and black negative values. The coils induce a rotating **E** field, that forces free charges to move. Current flow is high within the extracellular space, but the peaks occur in the gap junctions ($J_{\max}$ = 0.122 A/m^2), where the current flow densifies. From theory, one expects the highest amplitudes to occur underneath or at least close to the coil's ring. At locations where cells and **J** are not parallel high field values occur. Peaks appear at cell endings. Hence, the presented intersection plane can be divided in four symmetric quadrants: Along the symmetry axes, the **E** field is zero, whereas in the four quadrants of the plane, values $\neq 0$ arise within the cell membranes, depolarizing and hyperpolarizing the membrane accordingly.

Figures 4.18(c) and 4.18(d) show derived TMVs $E_x \cdot d$ and $E_y \cdot d$, respectively, with d being the membrane's thickness, extracted along the white dashed lines of the colored slice plots. Whether the **E** field is directed inward or outward of one cell is decisive for increasing or decreasing the natural resting potential.
Finally, the cumulative histograms presented in Figs. 4.18(e) and 4.18(f) summarize longitudinal and transverse positive TMVs occurring in the entire cell patch. 1 % of the cell membranes experience a lift of the TMV of at least 0.669 μV in longitudinal direction ($E_x \cdot d$), whereas 1 % of cell membranes experience a lift of the TMV of at least 1.162 μV in transverse direction.

Skeletal Muscle Tissue

Figure 4.19 presents sliced E_x and E_y, derived TMVs along a line through x and y and cumulative histograms resulting for replacing the myocardial cell model by the skeletal muscle patch. Due to the absence of intercellular connections, induced currents do not pass the membranes easily. The current within the cell membrane is much lower compared to extra- and intracellular currents. $J_{\text{max}} = 0.0255$ A/m^2 arises within the extracellular fluid.
The effect of increasing effective conductivity with frequency does not affect the amplitude of induced currents much. This has been shown in the human body simulations, too, where induced current densities rise linearly with stimulation frequency (Section 3.2.2.1). This implies that increasing the frequency facilitates cell excitation since all quantities scale with frequency. However, as will be shown later, excitability of cells decreases with rising stimulation frequency. Since the activation of ion channels within the cell membrane underlies specific time constants, cells do not react on stimulation impulses of short durations, leading to a higher threshold TMV.
In general, induced TMVs within the skeletal muscle patch are lower compared to TMVs in the myocardial model. This approves higher excitability of myocardial tissue due to its lower effective conductivity.

Single Nerve Cell Embedded in Muscle Tissue

Things are a little different as soon as a long nerve runs across the scene. The following Fig. 4.20 again presents the muscle model, but with a nerve fiber embedded. Figure 4.20(a) shows the magnitude of induced **E** field in logarithmic scale in order to emphasize the effect of the present nerve: The nerve's membrane presents highest values of induced **E** field. The conductivity of nerves is

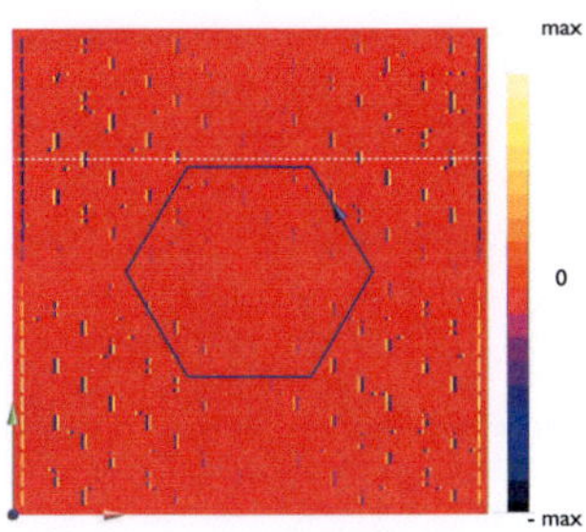

(a) E_x within the central x-y-plane of the myocardial model. The blue hexagon denotes the current loop, the white dashed line marks the extraction line for TMV derivation along x shown in (c).

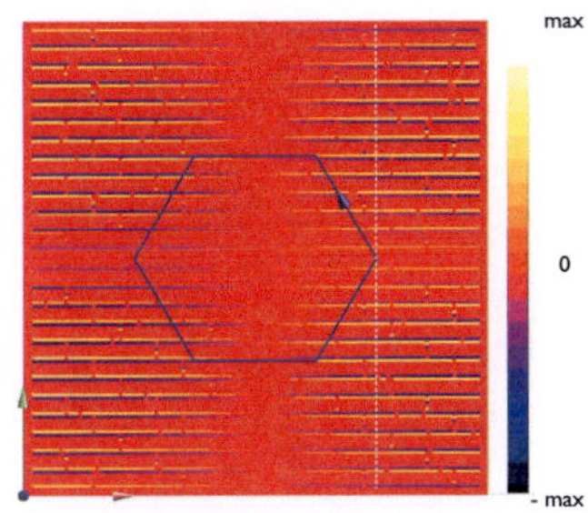

(b) E_y within the central x-y-plane of the myocardial model. The blue hexagon denotes the current loop, the white dashed line marks the extraction line for TMV derivation along y shown in (d).

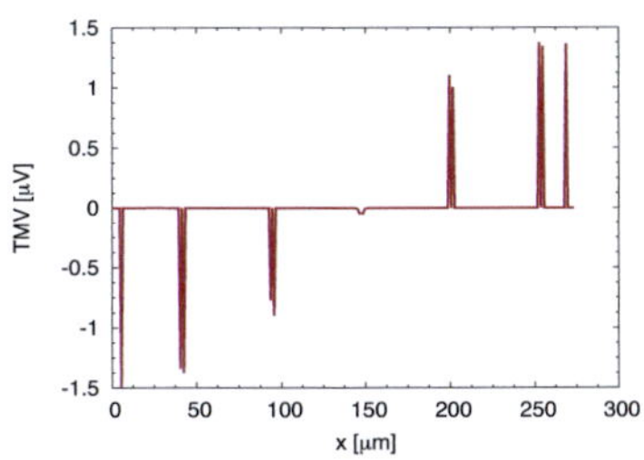

(c) Extracted TMVs along the line marked in (a).

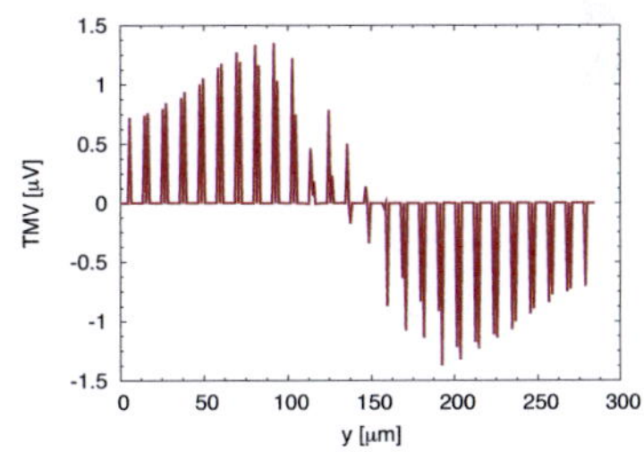

(d) Extracted TMVs along the line marked in (b).

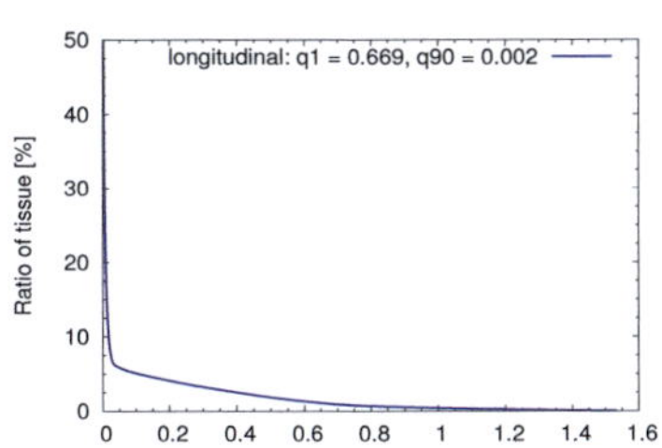

(e) Cumulative histograms of positive (depolarizing) TMVs caused byE_x.

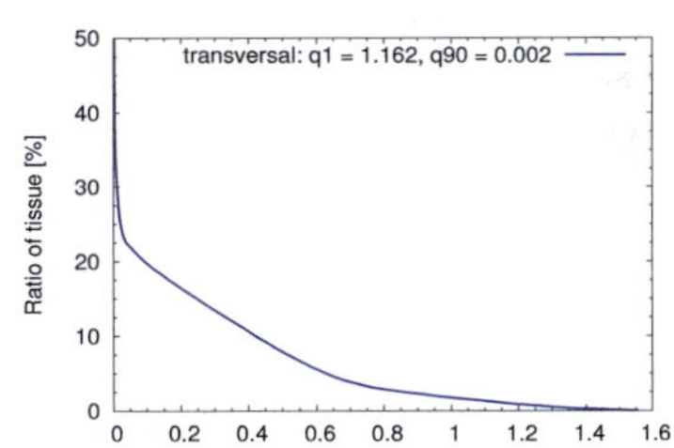

(f) Cumulative histograms of positive (depolarizing) TMVs caused by E_y

Fig. 4.18. E field components and derived TMVs induced in the myocardial cell patch model, penetrated by an oscillating **B** field of 16 mT and 10 kHz.

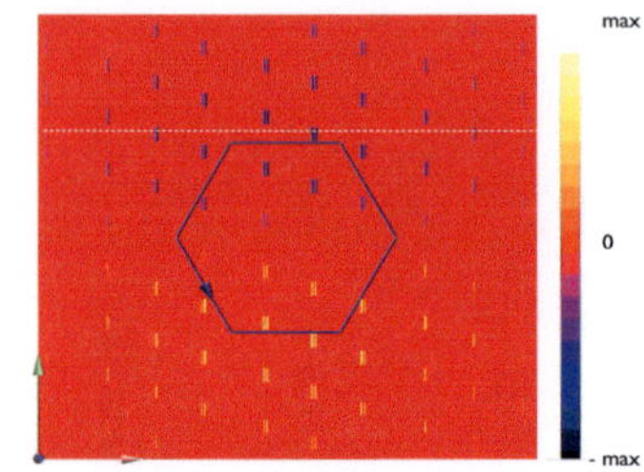

(a) E_x within the central x-y-plane of the skeletal model.

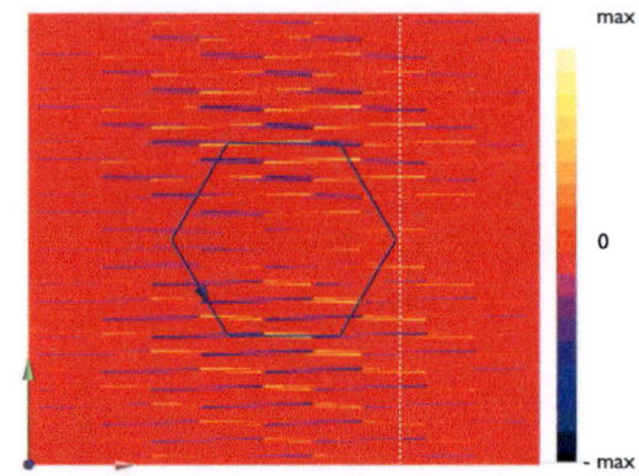

(b) E_y within the central x-y-plane of the skeletal model.

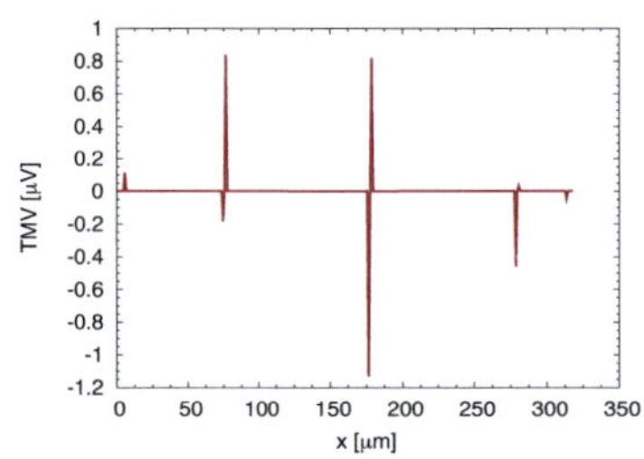

(c) Extracted TMV from (a).

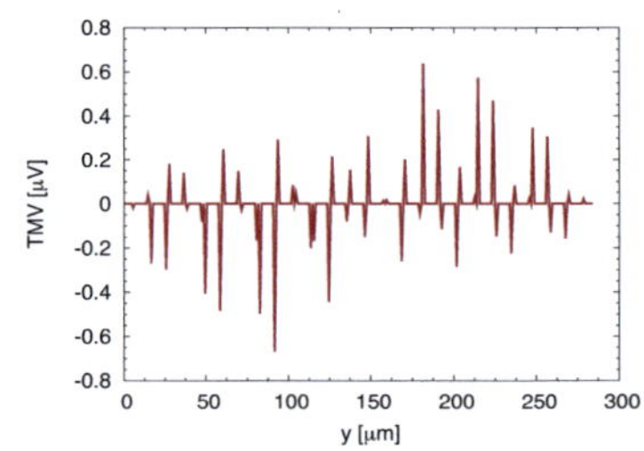

(d) Extracted TMV from (b).

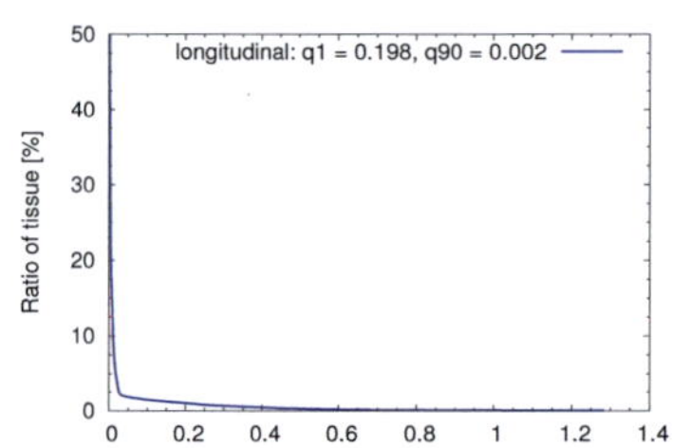

(e) Cumulative histograms of positive (depolarizing) TMVs caused by E_x

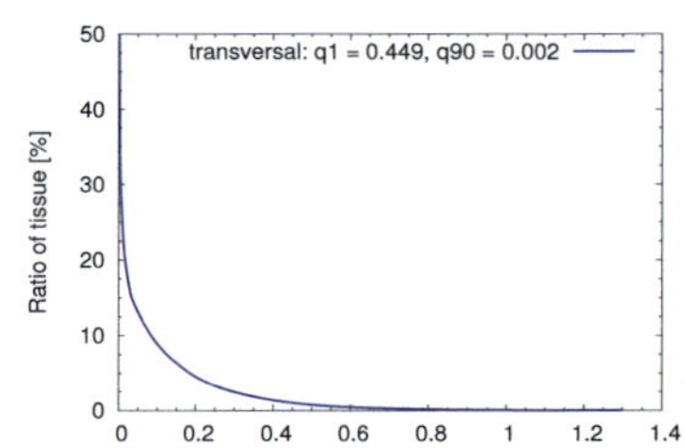

(f) Cumulative histograms of positive (depolarizing) TMVs caused by E_y

Fig. 4.19. **E** field components and derived TMVs induced in the skeletal muscle cell patch model, penetrated by an oscillating **B** field of 16 mT and 10 kHz.

much lower than both skeletal muscle and myocardial tissue. Hence, current distributions differ widely between inside and outside which leads to high TMVs. Figures 4.20(b) and 4.20(c) present the derived TMVs along the nerve's long side and transverse to it, respectively. Maximum values at the ends reach 2 μV, whereas only 1 % of muscular cell membranes do exceed 0.45 μV (see Fig.4.19(f)). Hence, the nerve would be the first to generate an action potential and would transfer the signal to its adjacent nerve or muscle cell by synaptic biochemical signal transfer. Table 4.10 summarizes induced field quantities in all three tissue models along with assumed threshold current density that was calculated again by scaling achieved quantities demanding a threshold TMV of 25 mV.

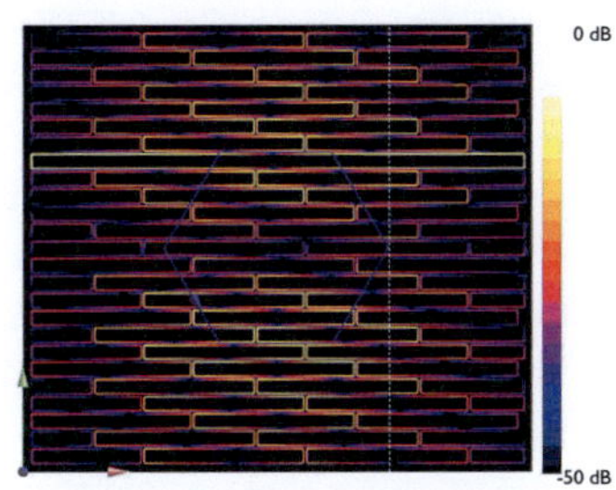

(a) Induced $|\mathbf{E}|$ (dB) in the skeletal muscle patch with a long nerve fiber embedded. The dB scale emphasizes the difference in induced field magnitudes.

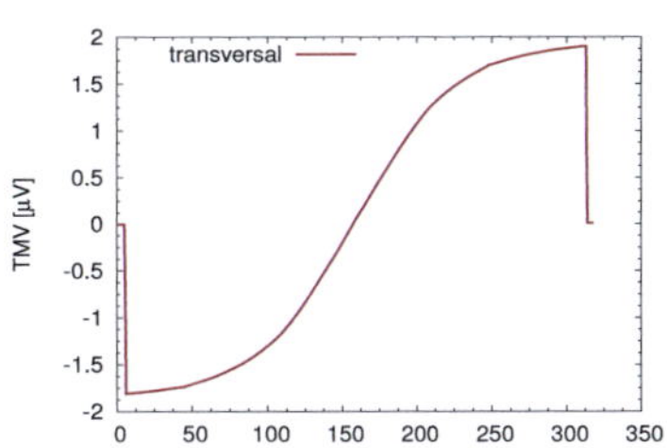

(b) TMV(x) = $E_y \cdot d$ extracted along the nerve's membrane

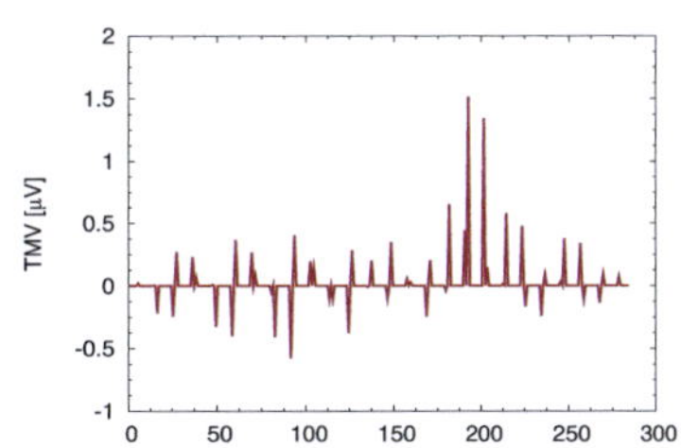

(c) TMV derived from E_y extracted along the dashed line in (a).

Fig. 4.20. Resulting TMVs in case of an embedded nerve in the skeletal muscle tissue

	q1 of longitudinal TMV [μV]	q1 of transverse TMV [μV]	J_{avg} [mA/m^2]	J_{thresh} [A/m^2]
myocardial model	0.669	1.162	0.766	27.8
skeletal muscle model	0.198	0.449	3.23	408
single nerve within muscle	1.9	–	0.323	4.25

Table 4.10. Extracted quantities from magnetic stimulations with 16 mT and 10 kHz. q1 values of TMV refer to the 1 % quantiles of extracted transmembrane voltages. J_{avg} is the average current density induced. J_{thresh} is the scaled J_{avg}. All values scale linearly with frequency.

4.3 Excitation Behavior of Dynamic Models

So far, the behavior of excitable tissues has been examined using numerical models that give information about field distributions depending on the source characteristics. The drawback of the numerical models is that they are static: Only the subthreshold behavior of the tissue can be calculated, providing that a steady stated is reached. Active behavior of nerves and muscles of course is not static. As is well known and described in Section 2.3.1, cell excitation depends on the time duration of the stimulus. Therefore, a dynamic model is required to provide information upon stimulation thresholds of excitable tissue.

In Section 2.3.2, electrophysiological models of myocardial cells have been introduced. In those models, every ion flow through the cell membrane is modeled by a set of differential equations. Usually, these models are employed for simulating natural healthy or pathological situations. They are not made for extreme conditions like fast stimulation, or high input currents. In order to make them usable for stimulation above 1 kHz, the calculation time step needed to be decreased and sinusoidal input signals were to be implemented.

The model developed by KIRSTEN H.W.J. TEN TUSSCHER and colleagues [36] was used here in order to investigate thresholds for action potential generation for sinusoidal stimulation above 1 kHz, with respect to frequency and stimulus duration.

The activator of action potential development is the sodium current. As described in Section 2.3.1, Na^+ streams into the cell after the threshold potential is reached. This Na^+ inflow is triggered by the opening of the respective ion gate in the cell membrane. As soon as this gate (in the model named *m-gate*), has reached the state "open", the generation of an action potential is triggered.

The input signal which disturbs the resting potential in the model is a normalized current injected into the intracellular space, given in pA/pF. Depending on the

amount of injected charge, the m-gate is forced to open or not. If the charge flow is too low, only very few ion-channels will reach the "open" state (1). Also, by injection of a "negative" current, the m-gate is forced to return to its resting state (closed: 0). Another observation that can be made is that even if the first period of the sinusoidal signal is not enough for stimulation, the second might be. This is due to the short time for the transmembrane voltage to return to its resting level. By the time of the second stimulus, V_m starts to be lifted from a higher V_0. All these effects imply that the shape and duration of the input signal, besides the amplitude, are decisive for action potential triggering. These effects have already been found in measurements in the 1960's [49,50].
In the following, threshold input values for action potential generation have been determined for frequencies between 1 kHz and 1 MHz. The duration of the sinusoidal stimulation was set to

- $t_p = T/2$ with $T = 1/f$
- $t_p = (n+1/2)T$
- $t_p = nT$
- $t_p = 100ms$

while $n \in N$. In the following Tables 4.11 and 4.12, amplitude I, number of periods n and the achieved $\Delta \mathrm{V_m}$ are given for first occurring action potential. $\Delta \mathrm{V_m}$ is given as amplitude of the reaction of the transmembrane voltage to the input signal. Q refers to the charge flow and is calculated by

$$Q = I \cdot t_p. \tag{4.21}$$

From Table 4.11, it seems that there is not much use in increasing n. Thresholds do not differ much for $t_p = T/2$ and $t_p = (n+1/2)T$. In spite of lower thresholds for increased stimulus duration, thresholds even increase for increasing n. Looking at the TMV, it seems to be drawn below resting potential by the regulatory system of gate opening and closing. With increasing frequency, higher currents are needed in order to reach the threshold potential. Furthermore, the threshold potential rises with increasing frequency, indicating that stimulation periods are too short, activation gates do not open.
Data given in Tables 4.11 and 4.12 are visualized in Fig. 4.21. Threshold amplitudes rise exponentially with frequency which goes along with theory. The stimulus duration does not seem to be of much effect. But whether the stimulus ends with a positive half sine, ($t = T/2$ and $t = (n+1/2)T$) or a negative half sine

	$t_p = T/2$			$t_p = (n+1/2)T$		
f	I	ΔV_m	Q	I	ΔV_m	n
[Hz]	[pA/pF]	[mV]	[nC/pF]	[pA/pF]	[mV]	
10	2.4	20	120	2.4	22	2
100	9	23	45	9.5	24	2
1000	80	26	40	81	25	2
5000	400	24	40	400	28	4
10000	800	26	40	1200	38	2
50000	3900	24	39	4000	26	2
100000	10700	34	53.5	7900	34	2
200000	25000	40	62.5			
400000	49000	40	61.25			

Table 4.11. Stimulation thresholds of the TEN TUSSCHER model for $t_p = T/2$ and $t_p = (n+1/2)T$

	$t_p = nT$			$t_p = 100\,ms$	
f	I	ΔV_m	n	I	ΔV_m
[Hz]	[pA/pF]	[mV]		[pA/pF]	[mV]
10	2.4	22	1	2.4	22
100	9.5	26	2	10	28
1000	95	32	2	90	30
5000	950	40	2	600	40
10000	1600	50	8000	1500	52
50000	11000	64	44000	11000	64
100000	33100	102	33100	22000	66

Table 4.12. Stimulation thresholds of the TEN TUSSCHER model for $t_p = nT$ and $t_p = 100\,ms$

($t = nT$ or $t = 100\,ms$) does affect threshold values. The last half sine pulls the TMV either up approaching the threshold potential or back down towards the resting potential.

Relating these results to the results from direct contact and magnetic stimulations (Sections 4.2.2 and 4.2.3), the achieved threshold TMV is the link. It is not the stimulation current, since in contrast to both magnetic stimulation and direct contact stimulation, the source current is injected into the extracellular space which alters the TMV. In Sections 4.2.2 and 4.2.3, threshold potentials were considered frequency independent. This is now proved wrong, as one can see now in the dynamic model. Threshold TMVs rise from 22 mV at 10 Hz up to 66 mV at 100 kHz, taking

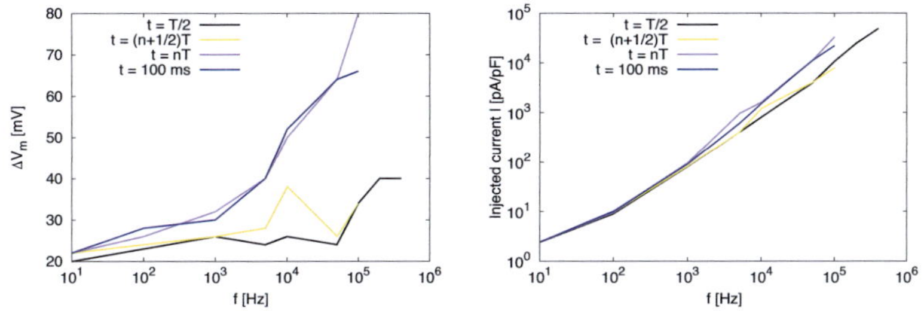

(a) Threshold potentials ΔV_m (b) Injected currents

Fig. 4.21. Threshold potentials and injected currents over frequency given in Tables 4.11 and 4.12

the results from the 100 ms stimulations approximated as "steady state" stimulation for relation to numerically obtained results. Taking this increase of threshold potential into account for direct contact and magnetic stimulations, threshold current densities from Tables 4.7, 4.8, 4.9 and 4.10 need to be re-calculated. There, it was assumed that 1 % of the membrane voltages needed to be raised by 25 mV. Corrected results induced averaged J for the myocardial tissue, skeletal muscle and neural tissue are given in Table 4.13.

f	TMV_{thresh}	J_{thresh} in heart		J_{thresh} in muscle		J_{thresh} in nerve	
		direct	magnetic	direct	magnetic	direct	magnetic
[Hz]	[mV]	[A/m^2]	[A/m^2]	[A/m^2]	[A/m^2]	[A/m^2]	[A/m^2]
1.0e+01	22	5.148	24.5	106	359	1.87	3.74
1.0e+02	28	16.46	31.1	139	457	7.95	4.76
1.0e+03	30	62.04	33.4	166	489	54.96	5.10
1.0e+04	52	366	57.8	788	848	566	8.83
2.5e+04	56	728	62.3	2,128	914	952	9.51
1.0e+05	66	3,432	73.4	6,864	1,090	2,503	11.21

Table 4.13. Re-calculated stimulation threshold current densities of excitable tissue models for direct and magnetic stimulation. Given values refer to tissue averages of induced current densities, when stimulating the tissue such that 1 % of the TMVs exceed frequency dependent threshold potentials.

4.4 Summary and Discussion

In this chapter, the stimulating effect of direct contact currents and time-varying magnetic fields on microscopic cell patches have been investigated. Therefore, voxel models representing myocardial tissue, skeletal muscle tissue and neural tissue have been generated, that vary in cell geometriy and configuration. Dielectric properties of the cell membrane and cellular fluids for each tissue representation have been determined based on well established data from literature.

Potential distributions of both direct and magnetic stimulation show realistic patterns, which reconstruct theoretical findings and published measurements. The magnitude of depolarization and hyperpolarization of cells besides input quantities depend on stimulating frequency and tissue characteristics. At low frequencies, depolarization takes place in the areas close to the stimulating electrodes (direct contact stimulation). The area of penetration gets smaller with increasing frequency. Instead, a regular saw-tooth pattern of TMVs is generated along the length of the cell patch, which implies synchronous cell excitation.

In magnetic stimulation, highest field induction takes place right below the stimulating coil. Cell membranes experience TMV alteration where the orientation of induced field relative to the cell orientation changes.

Relating induced current densities and transmembrane voltages, threshold current densities were determined for all three tissue types and frequencies up to 100 kHz, demanding a lift to a TMV above 25 mV in a volume fraction of 1 %. These current thresholds have finally been adjusted to frequency dependent threshold values of TMVs, that were determined by applying action potential generating stimuli to a well established dynamic cell model, which provides the course of TMV over time for sinusoidal injected current stimuli.

Current density thresholds throughout the frequency sweep are lowest for neural tissue, followed by myocardial tissue, while skeletal muscle tissue presents highest values. This is the expected order of threshold increase and reflects the natural behavior of the three different tissue types.

Threshold current densities of direct stimulation do rise stronger with frequency compared to J thresholds of magnetic stimulation. This results from differing stimulation mechanisms, which have been discussed in this chapter and explained in Section 2.4. REILLY predicts a pure linear relationship between stimulating currents and frequency above tissue characteristic bend frequency f_c, which originates from his simplifying assumptions of tissue and field homogeneities:

$$J = \sigma E_0 \cdot \frac{f}{f_c} \tag{4.22}$$

The results presented here provide more detailed results, since the models are not homogeneous and the threshold potential has not been regarded as frequency independent.
Comparing with measurements (see Table 2.5 in Section 2.4.2.2), evaluated results show a satisfying consent, despite the simplicity of cell patch structures.
After all, no comparable model has been found in literature that provides such detailed results of potential distribution patterns and threshold current densities.

5

Threshold Determination of Inductive Stimulation

"Nobody believes simulations". Everybody working with computer models sooner or later is confronted with statements like this. Simulations need to be validated and approved with support from a sensible experiment. In this regard, it needs to be proved that inductive stimulation is possible up to even 25 kHz. Hence, a system was built, that is capable of generating a magnetic field strong enough to induce a muscular reaction in humans. Necessary field amplitudes were recorded and labeled as "perception" threshold and "motor" threshold. Values varied quite much from subject to subject and not every subject sensed something. However, the experiment was successful in two ways: For MPI, it is a promising result, that the achieved thresholds are above drive field amplitudes, and from the experimental point of view, it was a success to show that magnetic stimulation up to 25 kHz is indeed possible.
In the following sections, the realization process will be described containing system specifications as well as hardware related considerations. After having the machinery ready to run, 31 subjects volunteered for participating in the experiment. The procedures of threshold determination and measurement evaluation are described in the Sections 5.4 and 5.5. Finally, the results will be interpreted and discussed.

5.1 Preliminary Considerations

The drive field coils in MPI generate an oscillating magnetic field of amplitudes between 10 mT and 30 mT which penetrates the whole body of the patient throughout the imaging procedure. Whether the drive fields are able to stimulate peripheral nerves or skeletal muscles might be found out by exposing superficial nerves and muscles to a magnetic field of MPI drive field amplitude and frequency. Results could underline simulation results. In order to localize cause and effect of

inductive stimulation, the idea was to generate a highly localized magnetic field strength which allows free choice of exposure position. Free setting of frequency, amplitude and stimulation time also were required.
Before the system was being built, some considerations preceded. Following topics had to be covered:

- expected stimulation threshold values, and therefore
- necessary coil currents in order to achieve and exceed those thresholds
- coil design
- circuit design that allows high currents flowing through the field generating coils

These considerations will be presented in the following, which led to the final stimulation setup.

5.1.1 Stimulation Site

When thinking about an adequate location for magnetic stimulation, several aspects need to be considered:

- The location must be easily accessible, so that shape or size of the stimulator do not restrain free positioning.
- In order to achieve super-threshold current densities, current pathways must be focussed.
- The stimulation success must be visible to guarantee an objective result.

With these aspects, the inner side of the upper arm, right above the elbow joint, was regarded as suitable site for magnetic stimulation. There, several muscles and nerves are located that control forearm and hand. The muscles however are probably too big to excite by stimulating a very localized spot. Yet, the ulnar nerve presents a promising excitation site. The ulnar nerve (see Fig. 5.1) is the largest nerve in the human body, that is not protected by covering muscles or bone. It is directly connected to the little finger, and the adjacent half of the ring finger, supplying the palmar side of these fingers, including both front and back of the tips.
These characteristics were considered convincing arguments for stimulation site selection. While the target arm area is easily accessible, the arm may rest in a carrier, allowing for a relaxed hand position.

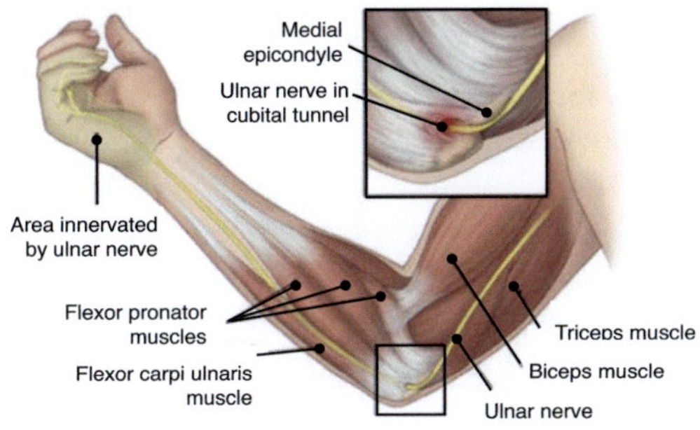

Fig. 5.1. Anatomy of the human arm [77]

5.2 System Specifications

5.2.1 Required Magnetic Field Characteristics

In order to build an appropriate field generating system that is potential to stimulate peripheral nerves and muscles, it is helpful to know the order of magnitude of the magnetic field amplitude necessary for tissue excitation. Findings from J. P. REILLY [71] as well as stimulation data summarized in the ICNIRP guidelines [60] had been taken into account. (See Section 2.4.2 for details.) From their considerations, induced current densities which exceed $0.1 \cdot f$ mA/m^2 (effective value, f in kHz) in the range between 1 kHz and 1 MHz are above stimulation thresholds. At 25 kHz, this results in a peak threshold current density of 3.56 A/m^2. The target magnitude of induced current density therefore was set to be some 10 A/m^2 to ensure stimulation success.

The second request to the field generating coil was to bundle the magnetic field at one point in order to achieve high field values at a well localized spot. Butterfly coils for example generate a field maximum at the intersection of the loops. (However, this is true only for ideal, sufficiently thin coil windings.) There are even more sophisticated coil designs for highly focused magnetic stimulation [78, 79]. For the sake of simplicity, the simple butterfly shape has been chosen as adequate coil geometry for this stimulation experiment.

The field distribution of a butterfly coil cannot be calculated analytically. Thus, a model of a simple butterfly coil was built in *SEMCAD X* in order to calculate the generated magnetic flux density and furthermore the induced current density in the human arm. Like before, the Visible Man data set was used here, along

with GABRIEL tissue properties. Field effects grow linearly with field generating coil currents and frequency. Thus all field amplitudes can be derived from one simulation in order to estimate the required coil currents.
For this ideal coil profile simulation setup, two current loops were positioned close to the bended elbow of the human arm. A distance of 4 mm between the coils and arm surface was kept to leave some space for insulation. In this arm area, nerves and muscles responsible for finger controlling run subcutaneous. Positioning high **B** field amplitudes there, aimed at inducing current densities oriented along nerves and muscle fibers, which increases stimulation possibility. Figure 5.2 presents the simulation model.

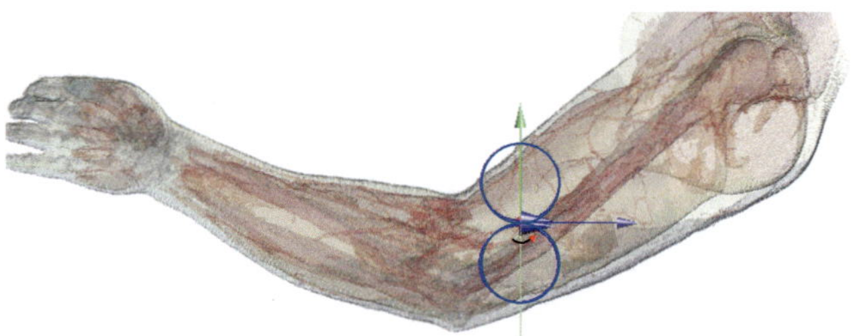

Fig. 5.2. Model in *SEMCAD X* of a simple butterfly coil setup. The red arrow marks the direction of the highest current density to be expected in the human arm model.

Running two current loops, each representing one wing of the butterfly coil, with $I = 1$ A at $f = 25$ kHz, the peak amplitude of magnetic flux density is 1.02 mT. Not surprisingly, the field maximum arises within the coils plane. Only 5.6 mm away in normal direction, $|\mathbf{B}|$ dropped to the half maximum value. Figure 5.3(a) shows the **B** field amplitude distribution in the arm's surface plane. The maximum induced current density in muscle tissue was 24.5 mA/m^2. Figure 5.3(b) presents a surface field view of current density amplitudes at the surface of muscle tissue. The factor between the achieved current density with 1 A coil current J_{1A} and the estimated stimulation threshold published by the ICNIRP J_{thresh} is

$$\frac{J_{1A}}{J_{thresh}} = cf = 6.88 \cdot 10^{-3}. \tag{5.1}$$

Thus, the coil current necessary to induce at least J_{thresh} in muscle tissue is $I/cf = 1/cfA = 145.3A$. In order to ensure muscle stimulation at 25 kHz, an

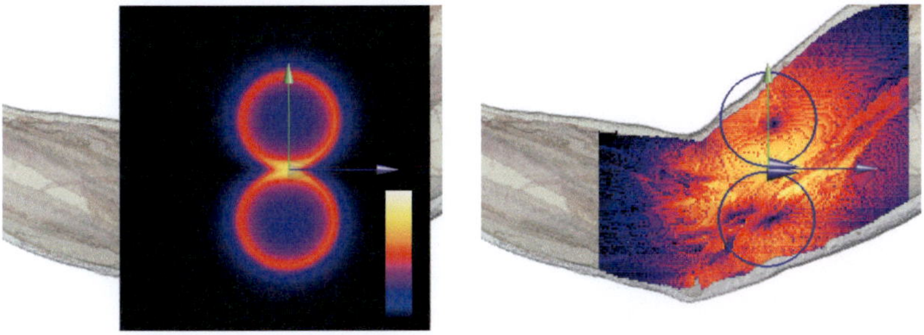

(a) Magnetic flux density, sliced at the muscle surface plane, normalized to the slice maximum of 62 μT

(b) Induced current density, surface field view, normalized to the tissue maximum of 24.5 mA/m^2

Fig. 5.3. Simulated fields generated by a simple butterfly coil with 1 A, 25 kHz.

extra factor of 10 was donated, which leads to a required coil current of $\approx$ 1450 A, or 1450 ampere-turns (AT), which may be used equivalently. All further considerations regarding system design were based on building a circuit that manages to generate a coil current of $\geq$ 1450 AT.

5.3 Hardware Realization

5.3.1 Circuit Design

In general, current magnification is achievable in a parallel resonant circuit, which - in ideal case - consists of an inductor L and capacitor C, connected in parallel. In this ideal resonant circuit, a frequency exists at which the energy oscillates between capacitor and inductor. This resonant frequency is defined by the components' properties:

$$f_r = \frac{1}{2\pi\sqrt{LC}} \tag{5.2}$$

The resonance effect occurs when inductive and capacitive reactances are equal in absolute value. No current flows through the input lead. If the applied frequency is below the resonant frequency f_r, the circuit represents an inductor. Above f_r, the circuit acts like a capacitor. As long as the resulting current is lower than the current flowing in the resonant circuit, current magnification is provided. In real parallel resonant circuits, the oscillation is damped by ohmic losses in the components. The capacitor's resistivity is negligible compared to the inductor's. The resistivity R of the inductor therefore is accounted for by serial connection to L. The equivalent resistivity Z which represents the input impedance in resonance reaches a maximum at:

$$Z = \frac{L}{R \cdot C} \tag{5.3}$$

The equivalent circuit is displayed in Fig. 5.4. Its resonant frequency is reduced due to the additional impedance. It is now given by:

$$f_r = \frac{1}{2\pi}\sqrt{\frac{1}{LC} - \frac{R^2}{L^2}} \tag{5.4}$$

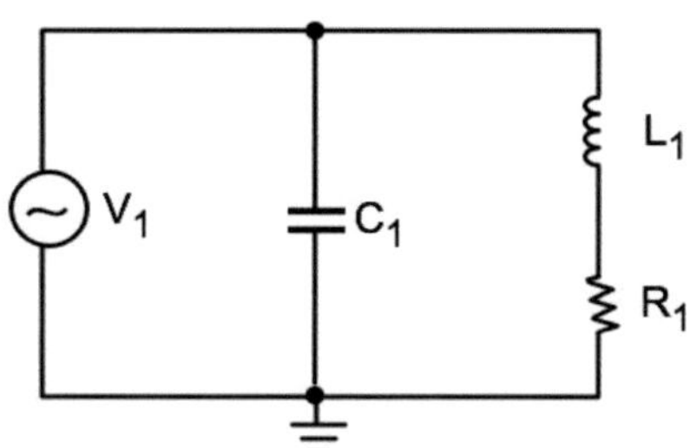

Fig. 5.4. Real parallel resonant circuit with R representing the coil's ohmic loss

Besides its resonant frequency and input impedance, a resonant circuit is characterized by its *Q-factor*, which is defined to be:

$$Q = \frac{f_r}{\Delta f} \tag{5.5}$$

The Q-factor represents the bandwidth, where the stored power is above 50 % of the maximum. The Q-factor of the parallel resonant circuit is given by:

$$Q = R \cdot \sqrt{\frac{C}{L}} \tag{5.6}$$

In the given scenario, L and R represent the inductance and resistivity of the field generating coil, respectively. The second constraints of circuit design was the resonant frequency. Consequently, since L and R were rather pre-determined, the only free parameter to trim the circuit, was the capacitor C. In order to manipulate f_r, C must be chosen accordingly. Increasing resonate frequencies can only be obtained by decreasing C. On the other hand, the input impedance rises with decreasing capacity (eq. (5.3)). If the input impedance gets too high, the current does not reach sufficient levels, despite very good Q-factors.

At this point, an additional capacity C_2 was inserted, connected in series to R and L (Fig. 5.5). Now, R, L and C_2 built a series resonant circuit. In resonance, this circuit provides voltage magnification which improves current draw. If C_2 is chosen to be significantly larger than C_1, the total C effective in the resonant frequency changes only little but the input impedance is reduced effectively. Thus a sufficient current magnification can be reached by appropriate selection of C_1 and C_2.

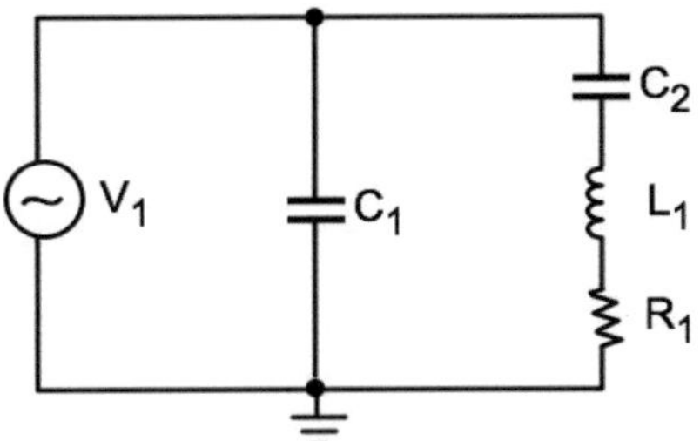

Fig. 5.5. Combination of a parallel resonant circuit with a series resonant circuit for impedance matching.

The circuit was simulated in a circuit simulation software *(SPICE: Simulation Program with Integrated Circuit Emphasis)* to estimate the system behavior prior to its manufacturing. Table 5.1 summarizes the results for the capacitors C_1 and C_2 for the applied approximations.

frequency [*kHz*]	required C [μF]	C_1 [μF]	C_2 [μF]
1	1180	1200	–
2.2	244	220	–
4.7	54	55	–
10	11	12	–
22	2.44	18	2.8
25	1.88	13.5	2.2

Table 5.1. Capacities for the parallel and series resonant circuit. Theoretically required total capacities C represent simulation results, C_1 and C_2 represent capacity separation for commercially available capacitors.

5.3.2 Coil Design

The field generating coil itself has an important system impact regarding its geometry: On the one hand, the generated field is proportional to the number of turns, since every single turn generates its own magnetic field which adds to the total magnetic field. This reduces the necessary coil current. On the other hand, the coil's inductance L rises with increasing number of coil windings, which in turn increases the input impedance of the circuit.
The coil geometry was chosen to approach the geometry of the *SEMCAD* model with 12 windings: 4 windings in 3 layers. For the preliminary *SPICE* simulation, L was estimated using the approximation of H. A. WHEELER[1] for short coils [80]:

$$L \approx \frac{\mu_0 N^2 \pi r^2}{l + 0.9r} \tag{5.7}$$

with N being the number of turns, r the medial radius and l the coil length, values in m. which resulted in $L_{1/2} = 10.87\,\mu H$ for one wing of the butterfly coil. The real coil's conductivity later turned out to be 21.4 μH, so the approximation of $L = 2L_{1/2}$ turned out to be quite a good estimate for circuit design simulations.
The coil's cable itself consisted of a high-frequency cord, built out of 2200 lacquer insulated filaments of 50 μm diameter each. The lacquer was approved to ensure a dielectric strength of 1 kV. Due to the expected voltage magnification in the series resonant circuit, the wire was additionally encased in 4 kV stable shrink sleeving.
The coil's housing consisted of rigid-PVC of 1 mm thickness. The winding cylinders had a diameter of 4 cm. In total, the housing fulfilled all necessary criteria regarding safety, mechanic stability, temperature stability and insulation. Furthermore, it was of a manageable size, offering flexibility in its handling and the possibility to position it close to its application site. Figure 5.6 presents the coil in its housing.

5.3.3 Capacitors

The parallel and serial capacitors C_1 and C_2 were realized by building capacitor arrays with regards to dielectric strength and required energy storage capacity. Air has a dielectric strength of $\approx$ 3.3 kV/mm, depending on temperature and humidity. Hence, every array was placed in its own housing, preventing accidental touching and assuring enough spatial separation between its components. Figure

[1] The original formula takes input measures in inches with μ_0 already included in the coefficients: $L = (a^2n^2)/(9a + 10b)\,\mu H$

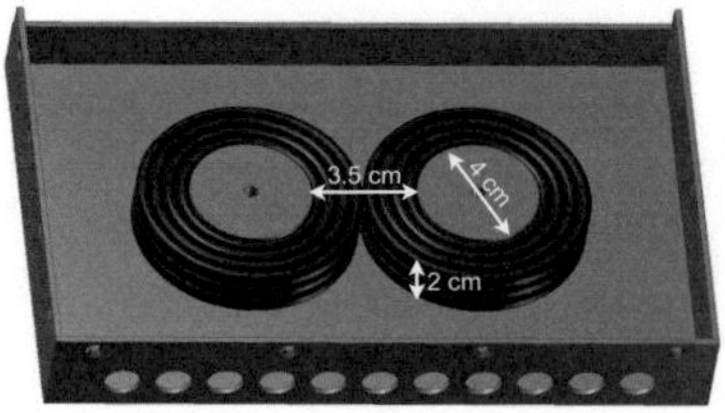

(a) CAD sketch

(b) Photo of the real coil

Fig. 5.6. Experimental field generating coil and housing

5.7 shows the capacitor array used as parallel capacitor C_1 in the 25 kHz circuit. It consists of 18 capacitor pairs connected in parallel. The capacitors[2] are cylindric, specified by their capacity of 1.5 μF, dielectric strength of 450 V and 12 A and a volume of 4.216 ml. Connecting 18 of those capacitors in pairs of two, the total capacity amounts to 13.5 μF, dielectric strength of 0.9 kV and 215 A_{eff} and total volume of 0.151 l. Typical polypropylene capacitors provide an energy density of 20 kJ/m^3. The energy in the circuit is estimated to be $E = \frac{1}{2}LI^2 = 0.154J$ (with I being the effective maximum current (Section 5.2.1) and L being the approximated coil inductance (Eq. (5.7))). With that, the energy density can be approximated to be $\frac{E}{V} = 1.02\frac{kJ}{m^3}$. Hence, a safety factor of ≈ 20 is provided. The other capacitors were selected accordingly, details given in Table 5.2. The number of circuits and thus the number of stimulation frequencies resulted from cost and feasibility considerations. Finally, four capacitor arrays were built, which enabled stimulation at four different frequencies due to multiple combination possibilities. After hardware completion, the frequencies were trimmed to the maximum achieved magnetic field generated by the coil.

5.3.4 Voltage Source and Amplifier

The stimulation circuit requires a considerably high input voltage. That was provided using a conventional frequency synthesizer together with the high-voltage amplifier DC-U 2250 from MEDTECH. It provides a maximum output voltage of 400 V (peak), maximum output current of 25 A (peak) and maximum power of 2.25 kW. Its internal resistance and inductance are given to be 12 Ω and 16 μH,

[2] All capacitors are from ARCOTRONICS

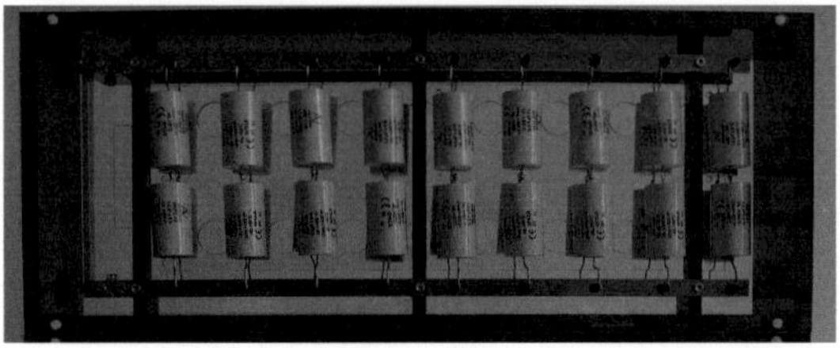

Fig. 5.7. Capacitor array of 20 capacitor pairs in its housing

frequency [*kHz*]	number of pairs	$C_{1:\text{single}}$ [μF]	C_1 [μF]	number of pairs	C_{single} [μF]	C_2 [μF]
7.4	18	2.5	22.5	–	–	–
12.04	18	1.5	13.5	18	2.5	22.5
16.5	18	1.5	13.5	14	0.22	+
				6	1.5	6.4
25	18	1.5	13.5	20	0.22	2.2

Table 5.2. Stimulation frequencies and capacitor assembly. C_1 represents the capacity of the parallel resonant circuit, C_2 the one of the series circuit.

respectively. Between its housing and low-lead, a capacitance of 10 nF is effective which disables potential separation with an isolating transformer. That's why every chance of touching circuit components had to be prevented. The DC-U 2250 amplifies incoming voltages by a factor of 100. Its current limiting is 28 A, which is only available for a very short time, since the thermal protective mechanism switches off the amplifier immediately. The generator was connected to the circuit via cable-end sleeves with screwable clamps.

5.3.5 Circuit Body, Safety and Comfort

The stimulation circuit was planned and implemented with much regard for user safety. Voltages and currents of such magnitudes as they show up in the circuit are of hazardous level. The possibility of accidental touching any charged component or current leading wire was reduced to zero. Switches between different resonant circuits were hidden in a box, which was placed at a secure distance to the subjects. Capacitor components were of high dielectric strength and spatially separated from each other. The butterfly coil was hidden in a case of 1 mm rigid-PVC guaranteeing sufficient dielectric strength and mechanical stability. While being fixed in its carrier, the coil was moveable in three degrees of freedom in order to bring it into optimal position. For the volunteers' arm, a fixture was built offering the most possible comfort, ensuring a relaxed arm position for the course of the stimulation experiment. The target position of field penetration was the same as in field simulation (see Fig. 5.2). Successful stimulation there may lead to a tickle, twitch or jerk either at the stimulation site or anywhere in the lower arm, wrist or fingers. All of these reactions were expected to be rather faint, neither pain nor harm in any way was intended.

5.4 Trial Proceeding

After the circuit had been finalized and the arm fixture was constructed, the hardware was ready to be used. Several items were yet on due:

System calibration: The simulated resonant frequencies were not really the ones expected in the real setup due to systematic errors like approximations in system characteristics (e. g. conductance L), tolerances in circuit elements (e. g. capacitor elements) and neglected properties (e. g. cable losses). Hence, each circuit needed to be trimmed to determine their actual respective f_r. Second, the relationship between input voltage and output magnetic field for stimulation was to investigate.

Trial validation: Since the results of the planned experiment highly depended on the subjective statement of each volunteer, trial validity somehow needed to be assured. Therefore, a preliminary trial upon 13 individuals was carried out in order to evaluate the significance of the subjective results.

5.4.1 System Calibration

After circuit completion, the final frequencies for the stimulation experiments had been settled and the point of maximum magnetic field strength was determined. The latter was achieved by measuring the generated magnetic field at discrete positions on the surface plain of the coil's housing. Therefore, pulse trains of low amplitude were used to keep the applied power low. For field measurements, a magnetic field sensor (type 11941A, AGILENT) in combination with a spectrum analyzer (type 4396B, HEWLETT-PACKARD) was used. The sensor itself was calibrated using a HELMHOLTZ coil of known field characteristics.

In reality, the field maximum generated by the coil was not located exactly at the symmetry point of the butterfly coil. This was expected due to geometric inaccuracies. Furthermore, the calculated resonant frequencies did not match actual frequencies of highest field generation. This was expected due to typical system inaccuracies:

- The coil inductance used for the *SPICE* simulations was an estimate of considerable error.
- Every single capacitor has a tolerance which affects the total capacitance and thus f_r of its resonant circuits.
- Cable losses had not been regarded in the *SPICE* simulations.

Therefore, after localization of the maximum field location, the actual resonant frequencies were determined by tuning towards maximized generated magnetic fields.

Finally, a $B(U)$ curve was measured for each of the four resonant circuits. They are presented in Fig. 5.8(a). As can be seen, the linear growth of generated magnetic flux density is damped beyond specific "cut-off"-voltages. This is due to current limiting of the amplifier. Figure 5.8(b) shows respective $dB(U)/dt$ curves which give maximum values of achievable rates of change of $B(t)$. Still, it must be considered, that every frequency represents its own circuit. The level of current magnification was not equal in all circuits. Consequently, during stimulation experiments, the input voltage was kept in the linear area of the $B(U)$ curves, below the respective cut-off-voltage, since increasing of the voltage did not lead to higher magnetic fields.

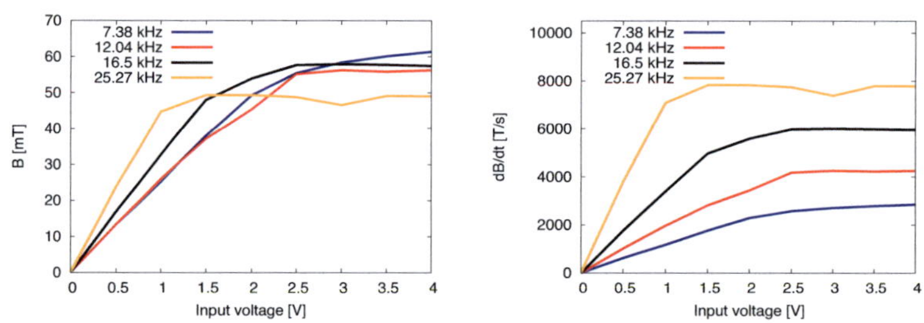

(a) $B(U)$ (b) $dB(U)dt$

Fig. 5.8. Realized maximum magnetic flux density B and rate of change dB/dt over input voltage U. The maximum achievable B and dB/dt are limited due to the amplifier's current limiting

5.4.2 Trial Validity Test

Statements about neural perception are very inaccurate and depend on the subject's sensation. Reaction upon magnetic field exposure is not measurable by any objective mean without high effort. Whether perception of an applied magnetic field is pure imagination or reality must be clearly determined in threshold determination studies. Everybody might feel a tickle at a specific spot in expectation to feel something right there. Furthermore, the sound of the amplifier and an audible click of hte coil in its housing when the machinery was turned on might lead to misleading results. To avoid misleading study results, the effect of the very focused magnetic field was studied in a separate group of volunteers.

Fig. 5.9. Trial setup: (1) voltage amplifier, (2) capacities C and C within their housing, (3) field generating coil within its housing (from [81])

The study proceeded as follows:

1. First, the 7.38 kHz circuit was connected, where the highest range of magnetic field was tunable. Pulse duration was set to 0.5 s, which is long enough for nerve excitation.
2. The subject's arm was fastened in the fixture, allowing to relax muscles of the forearm and fingers. The point of the coil's maximum field induction was positioned at the approximate target site of stimulation. Figure 5.9 displays the trial setup.

3. The first stimulation pulses were released in order to allocate the position of strongest visible reaction. If the volunteer sensed the first stimulation already, the arm position or coil position was further adjusted to the point of maximum sensation.
4. From there, the input voltages were decreased successively, until sensation vanished. The strength of lowest successful stimulation was recorded as stimulation threshold.
5. In order to approve the stimulation threshold, 20 stimuli were released, of which 10 were below threshold and 10 above in random order. After every pulse release, the subject was requested to classify the pulse as stimulative or not stimulative.

This preliminary study was carried out on 13 subjects, between 22 and 55 of age. The obtained thresholds were not included in the following experiment. 11 of 13 subjects passed the blank test with 100 % correct answers. 2 subjects had 2 false-positive statements, asserting that they sensed a sub-threshold stimulus. Supported by this success rate, the blank test was omitted in following trials in order to shorten study times.
Instead, the trials were run with regard to exposure time and stimulation frequency. After the optimal stimulation site was found, the exposure field was minimized to the perception threshold. Thereafter, the field was increased until some movement in one or more finger tips or in the heel of hand was visible. Thus, two thresholds, "perception" threshold and "motor" threshold, were recorded for each stimulation frequency and exposure time.
The first stimulation series was carried out at 7.38 kHz stimulation frequency, starting with an exposure time of 0.25 s. The exposure time was then successively shortened down to $n = 20$ oscillation periods.
In the second series the exposure time was kept constant at 0.25 s. By switching capacitor connections, the resonant frequency was increased without the need to move the arm.
The whole procedure (one subject) lasted between 30 and 45 minutes. 31 subjects, 22 male and 9 female, between 22 and 63 of age participated as volunteers.

5.5 Results of Investigated Magnetic Stimulation Thresholds

In the following, the obtained stimulation thresholds will be presented and discussed. Histograms and profile plots are used to extract significant information.

The y-axis in the histograms presents the number of events specified on the x-axis. The profile plot consists of n bins with each bin representing a subset of the observed data set. One bin presents the mean value μ of its subset and stretches from $\mu-\varepsilon$ to $\mu+\varepsilon$, with ε being the error of μ. The error is calculated by

$$\varepsilon=\frac{\sigma}{\sqrt{k}},$$

with σ being the standard deviation and k being the number of samples within the subset, respectively.

5.5.1 First Glance at the Data

Summarizing all data, 620 thresholds have been targeted. 182 trials were not successful, implying that the individual did not feel or react to the stimulus within the limits of applicable magnetic flux density. Fig. 5.10(a) visualizes the remaining 438 data points in a histogram which gives an overview of obtained magnetic stimulation thresholds. The overall mean is 50.41 mT while the standard error of this mean is 0.37 mT (below 1 %). The profile plot presented in Fig. 5.10(b) gives insight into individual results. Every bin represents one subject. Hence one bin takes a maximum of 20 threshold values into account (perception and motor reaction at 3 frequencies and 7 stimulation duration times at a fourth frequency), zeros excluded. The majority of the data is spread around 50 mT with few outliers (subject no. 2, 23 and 26). Those seem to be more sensible for magnetic stimulation. Most of the errors are rather small, suggesting that respective individuals' thresholds do not depend much on stimulation time or frequency.
In order to refine observations, the values need to be discussed separately, since they obtain different target events, namely thresholds for distinctive perception and thresholds for visible muscular reaction. Figure 5.11 presents the data in separate histograms for both sensor thresholds (Fig. 5.11(a)) and motor thresholds (Fig. 5.11(b)).
The distribution of perception thresholds ($B_{thresh:sens}$, Fig. 5.11(a)) presents its peak at 50 mT, with some sample counts around 20 mT. The distribution is not symmetric but has tails on both sides of the peak, in contrast to the distribution of motor thresholds ($B_{thresh:mot}$, Figure 5.11(a)). That one is shifted to the right, presenting a peak at 58 mT. No outliers towards lower **B** field values are evident. Those subjects showing low perception thresholds, all the same exhibit average motor thresholds. There's no tail towards higher B values, since generated **B** fields

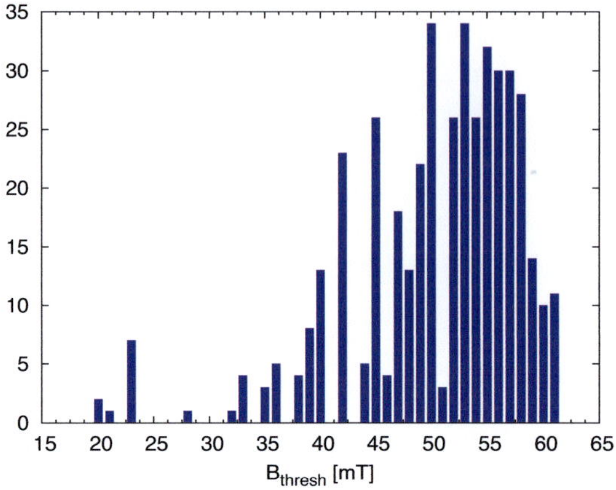

(a) Histogram of magnetic flux density thresholds for stimulation

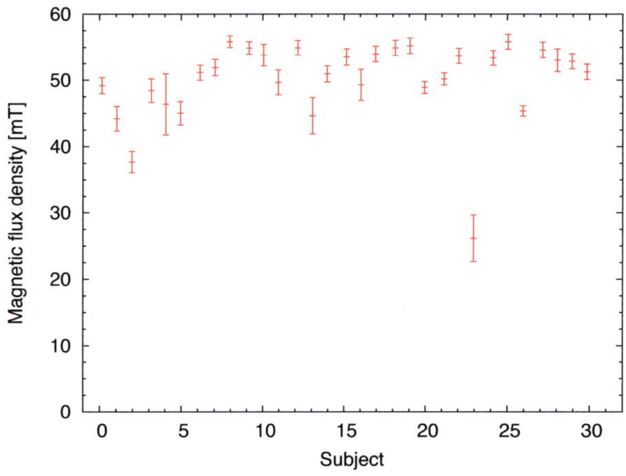

(b) Profile plot of all obtained B thresholds with respect to trial subjects

Fig. 5.10. Distributions of 438 successful threshold determinations with 31 subjects, including 4 frequencies. The overall mean value is 50.41 mT.

did not exceed 61 mT. Probably an increased **B** field would have provided a distribution similar to the one of perception thresholds; asymmetric, but with two tails instead of only one. In that case, results would appear as expected: Forcing a muscle to act takes a stronger input field compared to nervous reaction only. Fig. 5.12

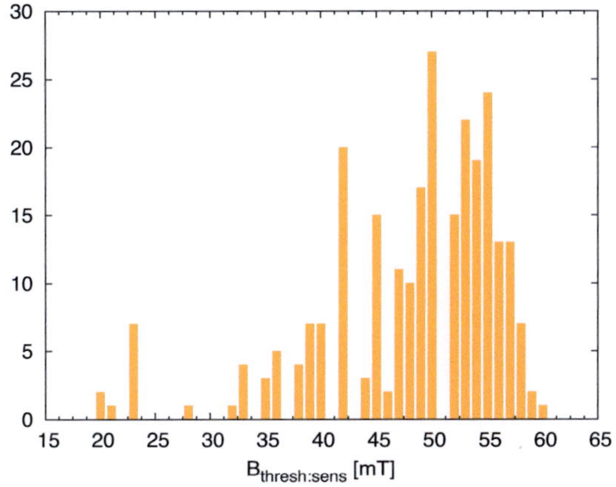

(a)

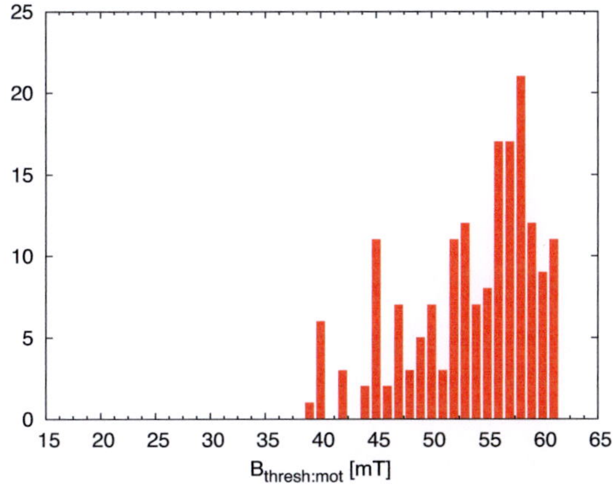

(b)

Fig. 5.11. Separated histograms of perception thresholds (a) and motor thresholds (b) approves this observation: The differences between corresponding motor thresholds and perception thresholds are given in a histogram. In most of the trials (25), the difference between motor threshold and perception threshold was below 0.5 mT or both values equaled. 55 % of the difference-samples are below 4.5 mT.

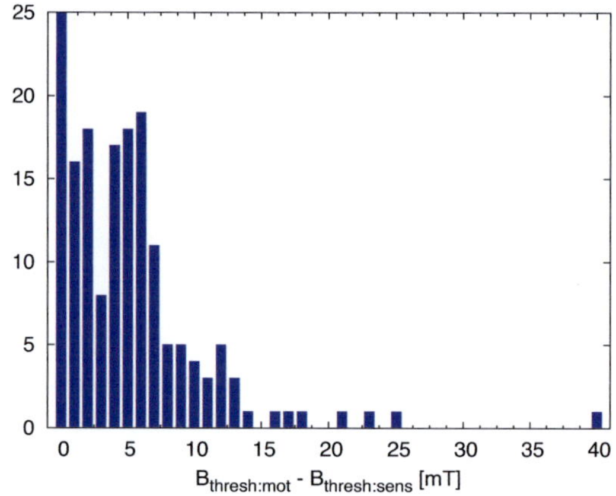

Fig. 5.12. Histogram view of differences between motor thresholds and perception thresholds

5.5.2 Stimulation Thresholds with Respect to Stimulation Times

Next, a more detailed view shall be provided. The results in figs. 5.13 and 5.14 represent perception thresholds and motor thresholds with respect to varying stimulation duration times. The stacked histogram in Fig. 5.13(a) presents the counts of obtained perception thresholds for every duration time. The peak summarizes thresholds between 50.5 mT and 51.5 mT counting 25 samples. Events of shorter duration times appear shifted towards higher *B* thresholds. This leads to an increasing number of failed stimulations due to **B** field limits. The trend towards higher necessary fields for successful stimulation can better be observed in the profile plot of Fig. 5.13(b). There, one bin represents all obtained thresholds at one specific stimulation time. The means of those decrease towards longer stimulation times, as expected from theory. The errors of that means change little over stimulation times.

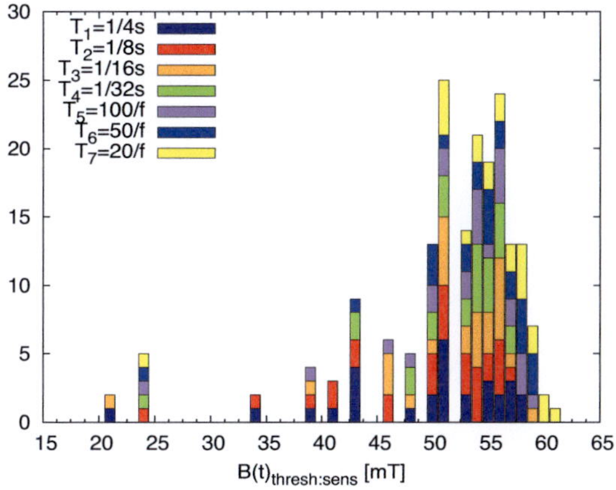

(a)

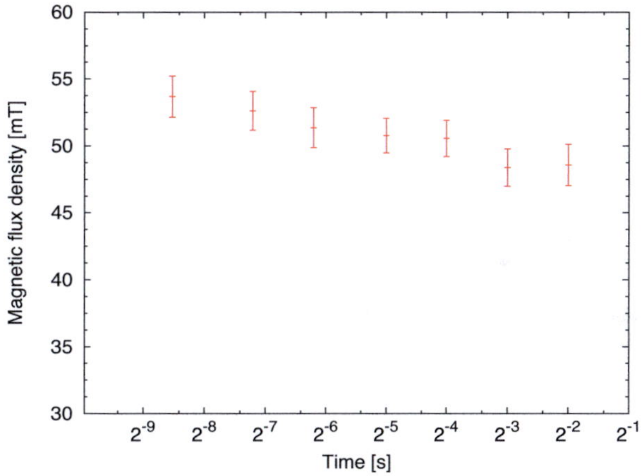

(b)

Fig. 5.13. Histogram and profile plot of perception thresholds with respect to stimulation duration times

Table 5.3 summarizes thresholds of magnetic flux density for perception at $f =$ 7.38 kHz. 29 out of 31 individuals reported perception of the stimulus at various spots down the forearm or at the stimulation site itself. One subject reported a tickle somewhere on the ascending path towards the shoulder. For pulse durations of 0.25 s, the average of those 29 recorded thresholds was 48.57 mT. The maximum to minimum threshold at which a doubtless field perception was assured, range from 57.18 mT down to 20.43 mT. Median values are slightly above mean values throughout all stimulations, which results from the asymmetric distribution.

T [s]	min(B) [mT]	max(B) [mT]	mean(B) [mT]	median(B) [mT]	max-min [mT]	count	ε [mT]
0.25	20.43	57.18	48.57	50.5	36.74	29	1.53
0.1250	22.81	56.0	48.45	50.5	33.19	31	1.39
0.0625	20.43	58.35	50.55	52.95	37.92	29	1.35
0.0312	22.81	56.0	50.76	52.95	33.19	27	1.29
0.0135	22.81	57.76	51.35	52.95	34.96	25	1.49
0.0068	22.81	58.35	52.6	0 54.18	35.55	25	1.44
0.0027	22.81	60.06	53.67	55.41	37.25	23	1.53

Table 5.3. Summary of **B** field thresholds for perception at $f = 7.38$ kHz. Only successful stimulations are taken into account, the number given in column *count*. ε refers to the statistical error of the mean.

Motor reactions varied from twitches in the fifth fingers to bendings of the whole wrist which led to whole hand lifts by few millimeters.

Fig. 5.14(a) presents the counts of obtained motor thresholds for every stimulation duration time. Compared to the distribution of perception thresholds, this is shifted towards the limit of applicable **B** field. With respect to pulse durations, a shift towards higher **B** fields does not seem to be significant. Looking at the profile plot in Fig. 5.14(b), one can see, that the incline of the means is not always negative towards longer stimulation times. Furthermore, the errors increase towards very short stimulation times of $n = 20$ oscillations or $T = 0.0027\,s$ of sinusoidal input signal. This reflects the decreasing number of successful motor reactions.

T [s]	min(B) [mT]	max(B) [mT]	mean(B) [mT]	median(B) [mT]	max-min [mT]	count	ε [mT]
0.25	44.75	61.06	55.27	56.59	16.31	24	1.03
0.1250	44.75	61.32	54.72	56.0	16.56	24	0.98
0.0625	44.75	60.81	55.69	56.59	16.06	21	0.87
0.0312	44.75	60.81	55.35	56.0	16.06	20	0.79
0.0135	47.01	61.06	55.81	57.18	14.05	17	0.90
0.0068	49.27	60.31	56.27	57.47	11.04	16	0.86
0.0027	51.72	61.32	57.21	57.47	9.591	8	1.29

Table 5.4. Motor thresholds of magnetic flux density amplitudes at 7.38 kHz

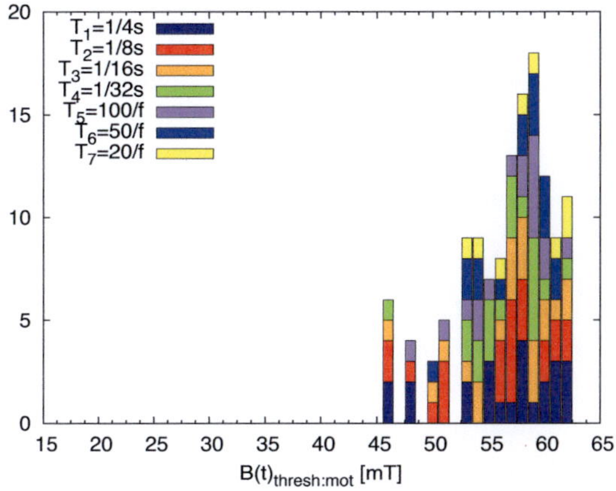

(a)

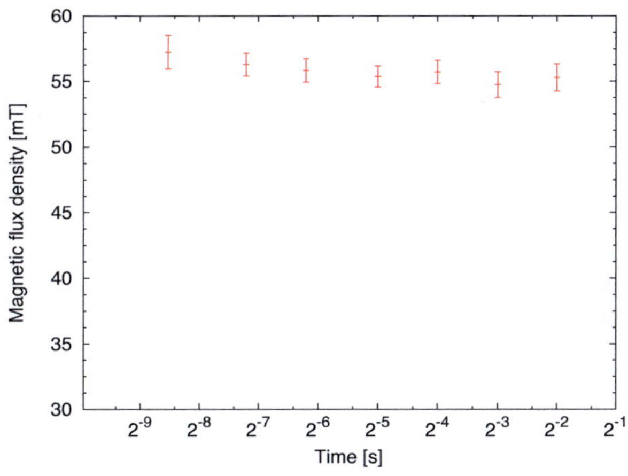

(b)

Fig. 5.14. Histogram and profile plot of motor thresholds with respect to stimulation duration times

5.5.3 Stimulation Thresholds with Respect to Stimulation Frequency

Threshold values regarding different field frequencies, keeping the stimulation time constant at 0.25 s will be presented and discussed accordingly. Figure 5.15 presents the histogram and profile plot of perception thresholds of both applied **B** and derived dB/dt. Looking at the histogram first, it is noteworthy, that there is one peak, stating that a **B** field amplitude around 42 mT is the perception threshold in 15 trials. It is also striking, that the distribution of B thresholds at 7.38 kHz are shifted towards higher B values, compared to the distributions of higher frequencies. However, it may not be concluded, that lower frequency requires higher fields for stimulation. Taking the rate of change of **B** into account, the events of different frequencies come into the right order (Fig. 5.15(b)). Thresholds of stimulations above 7.38 kHz do not exceed 52 mT. Regarding the histogram neglecting the different colors, one finds an almost symmetric distribution with an average value of 43.86 mT. At 7.38 kHz, 12.04 kHz and 16.05 kHz, the number of successful stimulations is almost constant, whereas at 25.27 kHz, only 17 out of 31 individuals sated definite perception within applicable **B** field limits.

The profile plot in Fig. 5.15(c) does not show a trend towards higher thresholds with increasing frequency. On the contrary, fig. 5.15(d) shows an almost linear increase. Still, the fact that the number of successful stimulation decreased to 17 at 25.27 kHz implies that stimulation becomes more and more improbable. Still, the error of the mean stays below 1.6 mT throughout all frequencies.

t [s]	min(B) [mT]	max(B) [mT]	mean(B) [mT]	median(B) [mT]	max-min [mT]	count	ε [mT]
7.38	20.43	57.18	48.57	50.5	0 36.74	29	1.53
12.04	20.83	49.22	40.09	41.19	28.39	30	1.16
16.5	23.22	52.75	44.1	44.91	29.53	28	1.22
25.27	27.93	49.28	42.7	45.56	21.35	17	1.59

Table 5.5. Perception thresholds of magnetic flux density amplitudes at various frequencies and 0.25 s pulse duration

Finally, determined motor thresholds are presented in Fig. 5.16 and Table 5.6. In the histogram view of threshold B amplitudes (Fig. 5.16(a)), it is no use talking about peaks and tails at all. Threshold counts of varying frequency spread wildly from 38 mT to 61 mT. The success rate drops from 24 at 7.38 kHz down to 5 at 25.27 kHz, increasing the error of the respective mean to 1.51 mT.

The profile plot in Fig. 5.16(c) shows the same characteristics as in Fig. 5.15(c), where perception thresholds are presented, but with increased average values. The increasing number of failing stimulations with increasing frequency might have two reasons: Rising improbability with increasing frequency and **B** field limits depending on the hardware. In all frequencies, the histogram shows samples at re-

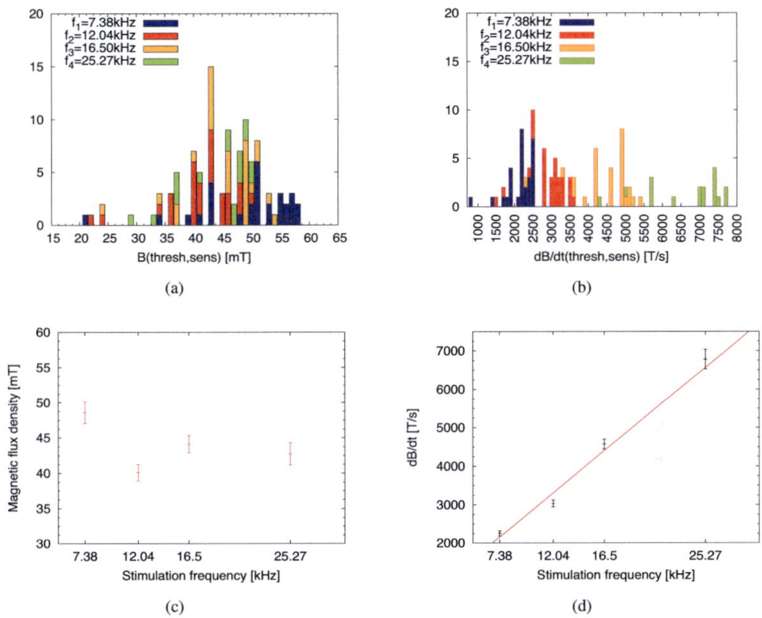

Fig. 5.15. Histogram and profile plot of perception thresholds of B and dB/dt with respect to stimulation frequencies

spective **B** field limits (see Fig. 5.8, where achievable **B** fields are given). This implies that there might have been more successful stimulations with higher thresholds, if those could have been generated by the coil.

t $[s]$	min(B) $[mT]$	max(B) $[mT]$	mean(B) $[mT]$	median(B) $[mT]$	max-min $[mT]$	count	ε $[mT]$
7.38	44.75	61.06	55.27	56.59	16.31	24	1.04
12.04	38.72	55.53	47.17	46.28	16.81	20	1.36
16.5	41.86	57.67	50.01	49.76	15.81	20	1.12
25.27	40.46	47.42	44.08	45.56	6.964	5	1.51

Table 5.6. Motor thresholds of magnetic flux density amplitudes at various frequencies and 0.25 s pulse duration

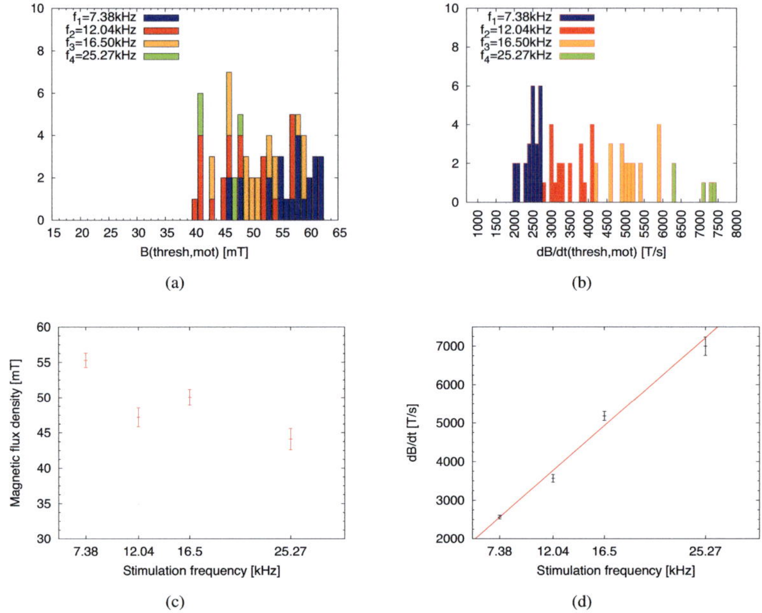

Fig. 5.16. Histogram and profile plot of motor thresholds of B and dB/dt with respect to stimulation frequencies

5.5.4 Stimulation Thresholds with Respect to Arm Circumference

In all those observations made in the previous sections, every individual was treated equal regardless of age, gender or shape. However, the anatomy of the arm might be the criterion with the highest impact on stimulation thresholds, since the magnetic field decreases very fast with increasing distance between the coil and stimulation site. Unfortunately, the statistical population is too small to investigate obtained thresholds with respect to the circumference of the individuals' arm *and* to frequency or time. Hence, positive stimulations have been observed all together. The bins in the profile plot of Fig. 5.17 present assembled thresholds of respective arm circumferences. Besides few outliers, a trend towards higher thresholds with increasing arm circumferences is visible. The line is fitted by using a χ^2-fit over all data. Its slope is 0.58 mT/cm.

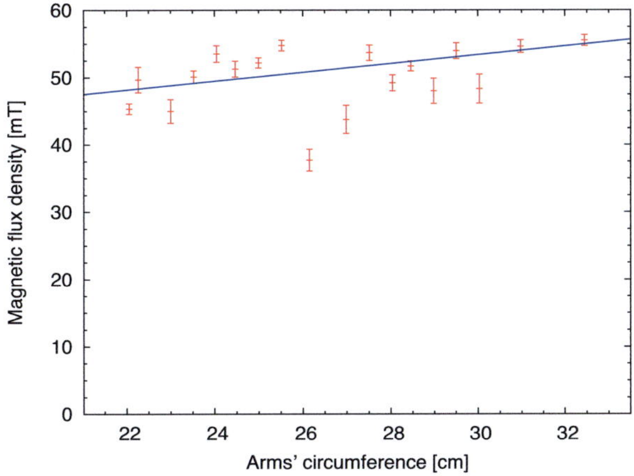

Fig. 5.17. Profile plot of thresholds with respect to arm circumferences. The slope of the χ^2-fitted line is 0.58 mT/cm.

5.6 Comparison with Simulated Data

Going back to field calculations, it is of interest to know the quantity of current density induced in excitable tissue of the arm by magnetic field densities in the mT range. In the simulation model, 12 current loops reconstruct one wing of the field generating coil, according to the real one. The distribution of the generated **B** field within the slice at the outside of the coil housing is illustrated in Fig. 5.18(a). The input current is $I = 100A$ in each coil turn. There are two peaks on either side of the symmetry point of the slice amounting 30.7 mT. 1.5 mm from there, deep within the arm, the maximum **B** field value dropped to 27.2 mT. That is where, in the arm of the Visible Man, first voxels of muscle tissue appear.
The generated magnetic flux density rises linearly with input current and so do induced electric field strengths and current density amplitudes. The maximum electric field and current density induced by 100 A coil turn current and 25.27 kHz is 32.9 V/m and 8.07 A/m^2 in muscle tissue, respectively. Figure 5.18(b) presents a surface view of J on muscle tissue in the area of stimulation. In order to account for frequency dispersion of dielectric properties of biological tissue, every frequency has been simulated in order to get current density distributions at 100 A input current. As will be seen, induced J does not rise exactly linearly with f.

From there, induced current densities can be derived over all measured **B** field amplitudes applied during the experiment according to

$$J_{ind} = \frac{B_{applied}}{B_{sim}} \cdot J(f)_{sim}, \tag{5.8}$$

where B values refer to values at the location of maximum field, whereas J values are maximum muscle values. Transferring obtained stimulation threshold B-field averages into induced current density thresholds leads to current density thresholds listed in Tables 5.7 and 5.8.

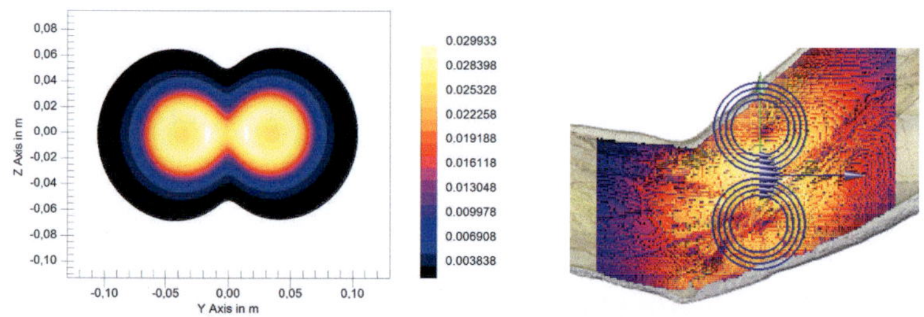

(a) B field distribution within the plain of the coil's housing. Note that there is no field beyond black areas.

(b) J distribution on the surface of muscle tissue. Maximum amplitude is 8.07 A/m^2 at 25.27 kHz

Fig. 5.18. Induced fields generated by 100 A in each coil turn.

	f	[kHz]	7.38	12.04	16.5	25.27
thresh:sens	mean(B)	[mT]	48.57	40.09	44.1	42.7
	mean($dBdt$)	[kT/s]	2.25	3.03	4.57	6.78
	J_{max}	[A/m^2]	3.65	4.96	7.51	11.22
	J_{mean}	[A/m^2]	0.297	0.404	0.611	0.912
thresh:mot	mean(B)	[mT]	55.27	47.17	50.01	44.08
	mean($dBdt$)	[kT/s]	2.56	3.57	5.18	7.0
	J_{max}	[A/m^2]	4.16	5.84	8.52	11.59
	J_{mean}	[A/m^2]	0.338	0.475	0.692	0.942

Table 5.7. Measured average B peaks and their corresponding simulated maximum and average current densities in muscle tissue, depending on applied field frequencies f

Figure 5.19 presents maximum and means of J thresholds from Tables 5.7 and 5.8 graphically. Thresholds seem to rise linearly with decreasing simulation times. The

	t_p	[s]	0.125	0.0625	0.0312	0.0135	0.0068	0.0027
thresh:sens	mean(B)	[mT]	48.45	50.55	50.76	51.35	52.6	53.67
	mean($dBdt$)	[kT/s]	2.25	2.34	2.35	2.38	2.44	2.50
	J_{max}	[A/m^2]	3.65	3.80	3.82	3.86	3.96	4.04
	J_{mean}	[A/m^2]	0.297	0.310	0.311	0.314	0.322	0.329
thresh:mot	mean(B)	[mT]	54.72	55.69	55.35	55.81	56.27	57.21
	mean($dBdt$)	[kT/s]	2.54	2.58	2.57	2.59	2.61	2.65
	J_{max}	[A/m^2]	4.12	4.19	4.16	4.20	4.23	4.30
	J_{mean}	[A/m^2]	0.335	0.341	0.339	0.342	0.345	0.350

Table 5.8. Measured average B peaks and their corresponding simulated maximum and average current densities in muscle tissue, depending on stimulation pulse durations T

frequency progress of induced J, appears to be linear, too, however the frequency band might be too narrow for approval. However, the trend towards increasing thresholds with increasing frequency is obvious.

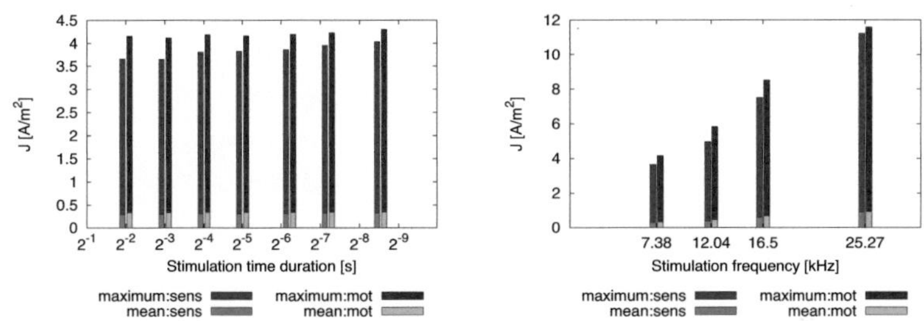

(a) Threshold J over stimulation times. Note the reversed x-axis. J increases with decreasing time durations.

(b) Threshold J over stimulation frequency

Fig. 5.19. Summarized threshold current densities for magnetic stimulation as result from the experiment combined with numerical simulations

5.7 Discussing Obtained Stimulation Thresholds

In this experiment, thresholds for magnetic stimulations have been investigated at four different frequencies and varying stimulation durations. Despite of simplicity of the setup, dispersion in the individuals' anatomy and sensitivity, results show a clear trend that goes along with theory.

After all, these results do not claim to be the one and only truth for excitation due to magnetic field exposure in the kHz range. Several aspects might be criticized which shall be discussed here:

- The value of measured **B** field that is referred to as B threshold, is the maximum **B** field measured with a magnetic field sensor. The sensor integrates **B** field lines, so that the actual peak might be even higher than values provided. This demur was erased by scanning the magnetic field over the surface plane of the coil's housing. The locations of the obtained B field peaks correspond well with the numerical simulations. Hence, those B values and locations are approved.
- These values are measured outside of the body. They decrease rapidly with increasing distance to the coil's plane. This has been taken into account in the simulations. However, derived current densities have a certain error, because of the anatomy of the arm model. As results have shown, the stimulation thresholds also vary with arm circumference, and so does induced J.
- In the experiment, it is obvious, that nervous excitation took place. It is much harder to stimulate muscles for several reasons. One reason is the size of affected tissue. The applied field in this trial was so much localized that the number of excited cells was not large enough to stimulate big muscles in the upper arm. Hence, the position of stimulation might have been not appropriate. However, this trial setup was created also with respect to accessibility, feasibility and comfort for the participants.
- Arguing with current density values in muscle tissue might be criticized to be inconsistent with reality. However, muscles are always connected to nerves. Trying to catch muscle tissue without affecting nerves magnetically, is rather improbable. Unfortunately, the Visible Human data set does not include peripheral nervous system. In that case, no assertions upon induced current densities in nerves can be made. But this is not the only uncertainty when using a numerical model: The model does not reflect an average human, it is only one sample. Furthermore, in this special simulation, the arm was not pressed against the housing of the stimulating coil, as the subjects' arms were. Hence the distance to excitable tissue was bigger in the simulation which leads to a smaller penetrated area. This of course influences calculated values of current density.

These results differ from threshold values determined by other authors. J. P. REILLY put the relation between dB/dt and frequency in the formula:

$$dB/dt_{\text{stim}} = dB_0/dt \cdot f/f_e, \tag{5.9}$$

with f_e being 3.35 kHz in case of peripheral nerve excitation and $dB_0/dt = 37.5\,T/s$ [46]. For f = 7.38 kHz, 12.04 kHz, 16.5 kHz and 25.27 kHz, threshold dB/dt peaks according to REILLY are 82.61 T/s , 134.8 T/s, 184.7 T/s and 282.9 T/s, respectively. The factors between experimentally determined thresholds here and theoretical values from REILLY, are respectively: 26.7, 22.5, 25.1 and 24.0 (= 24.50 in average). The difference might occur from the limitations of the system and the model discussed above. On the other hand, the theoretical model from REILLY also includes several simplifications which are described in Section 2.4.2. The most important difference is that the values obtained in this experiment are the result of a very much localized **B** field whereas in the studies of REILLY a homogeneous distribution is the basis for threshold considerations. Regarding B thresholds with respect to stimulation time, REILLY gives the following relationship:

$$dB/dt_{\text{stim}} = dB_0/dt \cdot \tau_e/\tau_p, \tag{5.10}$$

for $\tau_p \leq 0.149\,ms$, and constant otherwise. Stimulation times in the experiment are all longer than 0.149 ms and still thresholds vary with stimulation times (Fig. 5.19(a)).

Experimentally determined stimulation threshold current densities are within the range of earlier determined thresholds, summarized in Table 2.5, although comparison is difficult due to different trial settings.

As for current density restrictions, determined J values are above assumed thresholds by the ICNIRP. From the formulation $J_{\text{thresh:eff}} = f/10\,A/m^2$, f in Hz, thresholds result to be 1.04 A/m^2, 1.70 A/m^2, 2.33 A/m^2 and 3.57 A/m^2 for the given frequencies. The factors between experimental values and ICNIRP values are 3.5, 2.9, 3.2 and 3.1 (= 3.19 in average). The ICNIRP refers to the first reactions that have ever been measured, hence this deviation of thresholds is quite plausible.

In summary, it is concluded, that the measured B values are trustworthy in every respect. Derived current density thresholds may have a quite large error but are the best that can be obtained today.

Part III

Closing Matters

6

Summary

This project was started as part of the joint research project *MAGIC* of the German Federal Ministry of Education and Research (BMBF), conducted by PHILIPS RESEARCH, Hamburg, Sector Medical Imaging Systems. With the invention of Magnetic Particle Imaging, the need arose to learn about excitation mechanisms in the kHz range and potential tissue warming. Various projects have been carried out in order to investigate the effects of time-varying magnetic fields on the human body at frequencies between 1 kHz and 100 kHz. The linkage between different projects is one challenge this research had to cope with. Simulating induced current densities within a numerical body model is one task, evaluating the capability of stimulation another. To make things even worse, theory and practice hardly ever fit together. This chapter summarizes and discusses achieved results and finally ends with an outlook towards further necessary research.

6.1 Discussion of Methods and Results

Using a numerical body model provides the opportunity to look into the body in a macroscopic way. Organs are treated as homogeneous entities which can be assigned frequency dependent dielectric and thermal properties. Field quantities are provided for every location x, y and z. The limiting factor is the size of the grid which determines the degree of freedoms of the numerical problem. Combining the stage of development of the software *SEMCAD X* and the available hardware, the upper limit of grid nodes was reached at about 40 million in magneto quasi-static problems and 20 million in the electro quasi-static environment. The human body model used was resolved in such detail that the limit of computable nodes was reached. The choice of the computational grid aimed at high resolution within the core body in order to achieve reasonable results even at tissue boundaries. The most critical organ considering stimulation is the heart since its stimulation may lead to hazardous premature beats or even cause flutter or fibrillation. In favor of high torso resolution, the limbs have been neglected completely in the numerical simulations. Probably, induced field amplitudes are even higher within the arms but on the other hand are more tolerable. The head has been left out throughout all considerations. The available model does not provide necessary details about the brains and eyes, which are critical organs with respect to both heat and stimulation. Furthermore, the knowledge in brain electrophysiology is not sufficient for investigation of field effects on the human brain.

6.1.1 Calculation of Field Distributions

By means of numerical calculations upon a human body model, amplitudes of induced current densities, Specific Absorption Rates and temperature increase have been calculated given that the central magnetic flux density is 10 mT and the exposure time is 5 minutes. Simulation frequencies were 10 kHz, 25 kHz, 50 kHz and 100 kHz (Chapter 3). 25 kHz is the frequency planned for MPI drive field coils. It was shown that temperature increase even below 100 kHz is not negligible. SAR restrictions, which are supposed to limit thermal heating, are not defined for frequencies below 100 kHz. For the course of 5 min, temperature increase at 25 kHz was estimated to be around 1.5 K at the maximum, whereas 99 % of the total torso do not exceed 0.6 K, which is a tolerable value. While applying the PENNES bio-heat equation, a worst case scenario was constructed. The human body's thermoregulatory system is probably capable of handling higher temperatures than those given in the restrictions. Furthermore, the duration of 5 minutes is a rough estimate of planned MPI duration time. Developers claim that the imaging procedure will be much shorter. After all, induced temperature rising stays below hazardous levels for frequencies up to 25 kHz, although calculated local SAR values do exceed thresholds defined for Magnetic Resonance Imaging.

As for current density distributions, the effects of those is difficult to classify. Restrictions are based on measured values from direct electrode stimulations. However, those measurements consider the current flowing within the leads, distributed over the electrode surface area. The real effective stimulating current in tissue is literally impossible to assess. Hence, field restrictions include an uncertainty factor in addition to the safety factor that further reduces the risk of unwanted nerve and muscle stimulation.

Calculated maximum current density values by far exceed restrictions for induced current densities. In case of 25 kHz, the limit for occupational exposure is 0.365

A/m^2, averaged over 1 cm^2 of affected area. Maximum calculated values exceed 100 A/m^2 in muscle tissue and 40 A/m^2 in the heart, while averaged values are much lower (less than 25 A/m^2 in muscle tissue and below 10 A/m^2 in the heart). Calculated maximum values even rise with grid refinement. Here, the cumulative histograms give a more detailed idea of current distribution.

By superimposing three magnetic fields of slightly differing frequencies, as it is practiced in MPI for scanning the field of view, a beat frequency of 46.5 Hz envelopes the oscillations. At 46.5 Hz, stimulation thresholds are much lower than in the kHz range, since currents do not pass the cell membranes and hence high transmembrane voltages arise. Calculated q1 current densities after superposition of fields reach values of 108 A/m^2 in muscle tissue and 39 A/m^2 in the heart. Whether the patient feels the beat frequency or not, cannot be stated at this point.

6.1.2 Estimation of Thresholds for Heart, Muscle and Nerve Excitation

The stimulating effect of induced current densities has been approached to evaluate with microscopic models of patches of excitable tissues (Chapter 4). After adaption of the properties of main tissue components, those very schematic cell models have been employed for investigation of the membrane behavior upon direct and magnetic stimulation. Averaged conductivity and permittivity matched GABRIEL values well. Extracted transmembrane voltages for different cell geometries and configurations have been related to the respective current density distribution within tissue. Assuming that 1 % of depolarized membranes are enough for cell excitation, tissue threshold current densities for magnetic stimulation at 25 kHz are 9.51 A/m^2 for neural tissue, 62.3 A/m^2 for myocardial tissue and 914 A/m^2 in skeletal muscle tissue. These values make sense, implying that muscle stimulation happens only for very high fields, whereas neural tissue is the first to react on magnetic fields. which has been proved in the stimulation experiment described in Chapter 5. There it turned out, that stimulation of excitable tissue is possible at much lower current densities. The employed cell models describe well qualitative field distributions like current density and electric potentials. But the simplifications that needed to be made, required assumptions that seem to prevent proper quantification.

The stimulation experiment was carried out in order to underline simulation results. The necessary hardware was developed and assembled with regard to maximum achievable magnetic fields generated by a small figure-of-eight coil, with high effort for human comfort and safety. The magnetic field produced by the coil stimulated the ulnar nerve in the upper arm of 31 volunteers at four different frequencies and seven different stimulation times. In order to determine the simulating current density in the arm, the setup was reconstructed in the simulation setting. Current densities induced in the arm at field strengths that had been determined as perception and motor thresholds, exceed ICNIRP values for assumed thresholds by a factor of 4.

At 25 kHz, the maximum and volume average of current density induced by the averaged **B** for perception were 11.22 A/m^2 and 0.912 A/m^2, respectively, whereas the maximum and volume average of induced J for muscular reaction in the hand were 11.59 A/m^2 and 0.942 A/m^2. The cumulative histogram of induced J is visualized in Appendix A.1.5. The q1 value at 25 kHz is given there to be 4.883 A/m^2. The ICNIRP assume that stimulation occurs if $|J|$ is 3.57 A/m^2. This value is achieved or exceeded in 2.4 % of 1.5 l of arm muscle tissue ("q2.4 value").

Comparing these findings with simulated current densities in the human torso model, the q1 value (4.883 A/m^2) is exceeded in 65.3 % of peripheral muscle tissue, whereas the maximum $J = 11.59$ A/m^2 is reached or exceeded in 24 %. These considerations imply that patients will experience muscle twitches during MPI. Unfortunately, the human model does not contain a continuous web of nerves. From cell patch simulations, threshold current densities for muscle tissue are much higher than extracted values from the stimulation experiment. Muscle tissue will not be excited directly, but embedded nerves might be, which send sensor and motor signals to adjacent muscle, neural or neuronal cells.

6.2 Conclusions

6.2.1 Magnetic Stimulation

Numerical simulations are yet the best method available to estimate field distribution within complex human models. Calculated current densities induced in the torso model are within a reasonable range. Exposing humans to time-varying magnetic fields might cause nerve activation or muscle contraction at least up to 25 kHz. A wide spread belief, that stimulation is not possible above 5 kHz, was proved wrong.

Although the quantification of field distributions of numerical cell models turned out to be misleading, the obtained qualitative distribution patterns of excitable structure are reasonable. From these results, it can be seen, that membrane depolarization and hyperpolarization is evident up to 100 kHz.

6.2.2 Body Warming

The calculated temperature increase showed that even below 100 kHz, local tissue warming is not negligible. However, maximum values obtained may probably not be reached, since the lack of simulated blood flow presents a kind of worst case estimation. It is concluded that for MPI, temperature increase is not a limiting factor at 25 kHz but should be observed continuously when setting MPI drive field parameters. Increasing the drive field frequency reduces the risk of stimulation but increases heating effects.

6.3 Perspectives

Magnetic Particle Imaging is a promising imaging technique. The development is ongoing. To date, the construction of the first human scanner is scheduled within the scope of this decade. Parameters like coil geometries and settings in this thesis are the stage of development of 2010. In case drive field parameters change, results need to be adapted or re-calculated.

As long as computational power grows, numerical algorithms will be further developed. With that, larger matrices with more degrees of freedoms can be handled and more complex problems can be solved. Furthermore, more sophisticated human models are being generated, with more details in muscle structures and central nervous system. Limitations like missing fiber orientation in muscle tissue and the heart might get abolished and blood vessels accounting for heat transfer by blood flow might be included.

Furthermore, analytical nerve and neuronal models might help to investigate stimulation thresholds of single nerves, nerve bundles or neurons. Today already, magnetic stimulation is used in therapy, although the mechanisms are not yet fully understood.

A

Appendix

A.1 Human Body Simulations: Current Density Cumulative Histograms

In the following, cumulative histograms of $|\mathbf{J}|$ of excitable tissues are displayed. Cumulative histograms give more detailed information since they read the relative frequency of occurrence in tissue. Maximal values in one voxel in a specific tissue are prone to large stair-casing errors associated with sharp corners of the cubic voxel. A solution to obtain more stable peak approximations is based on choosing for the peak value a value representing the 99th percentile value of the induced field in a specific tissue [61]. From the biological point of view however, this is a somewhat arbitrary choice because the peak value depends on the grid resolution. The following graphs present induced $|\mathbf{J}|$ of every coil pair x, y and z. The legends also read the q1 and q90 values (which correspond to the 99th and 10th percentiles in ICNIRP nomenclatur) which give the maximum J occurring in 1 % and 90 % of tissue volume.

A.1.1 Simulations $f = 10$ kHz, $|\mathbf{B}|_{\text{center}} = 10$ mT

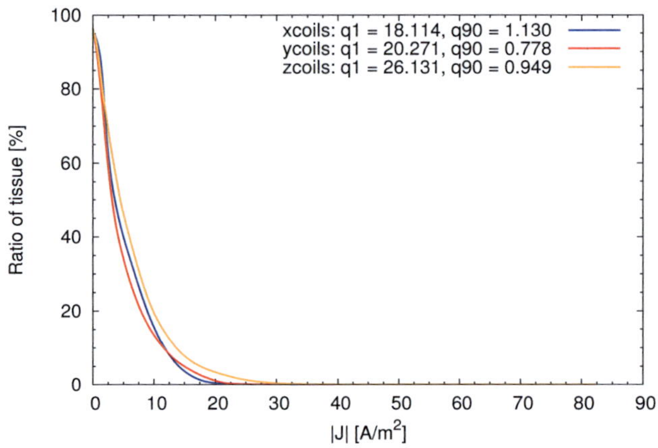

Fig. A.1. Cumulative histogram of induced $|J|$ at 10 kHz within the entire torso (total volume: 65 l)

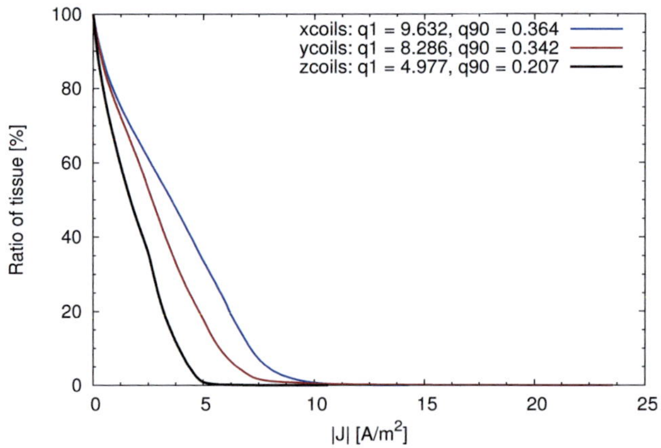

Fig. A.2. Cumulative histogram of induced $|J|$ at 10 kHz within the heart (total volume: 0.35 l)

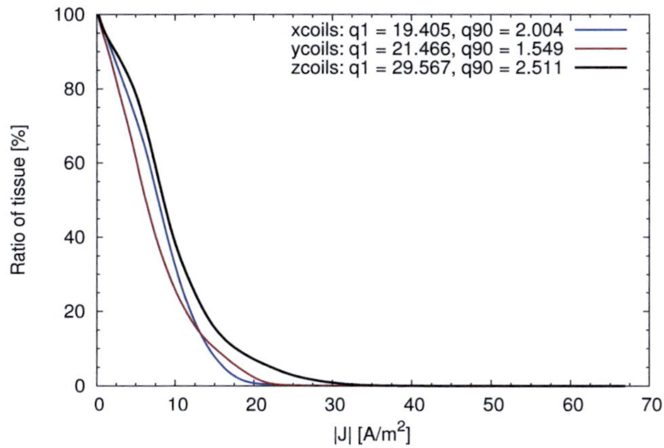

Fig. A.3. Cumulative histogram of induced $|J|$ at 10 kHz within muscle tissue (total volume: 23.2 l)

A.1.2 Simulations $f = 25$ kHz, $|\mathbf{B}|_{\text{center}} = 10$ mT

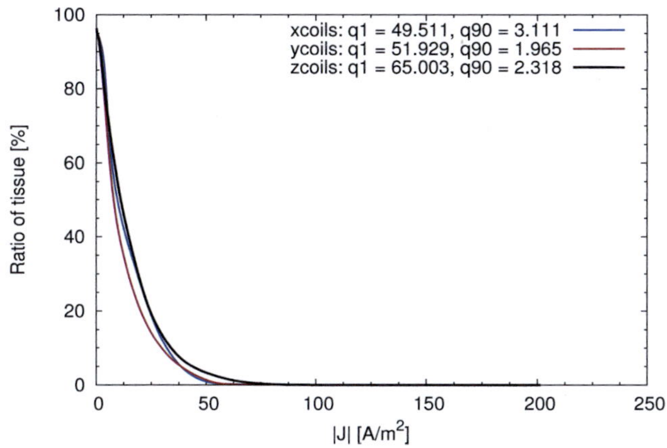

Fig. A.4. Cumulative histogram of induced $|J|$ at 25 kHz within the entire torso (total volume: 65 l)

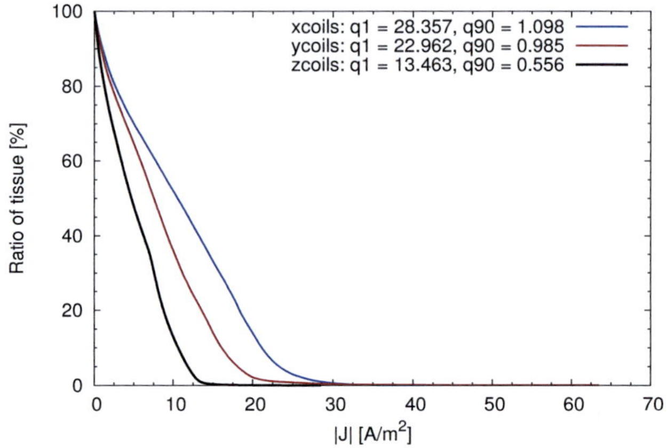

Fig. A.5. Cumulative histogram of induced $|J|$ at 25 kHz within the heart (total volume: 0.35 l)

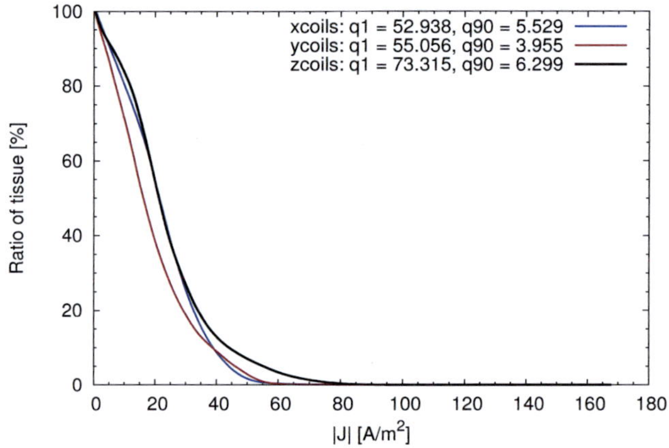

Fig. A.6. Cumulative histogram of induced $|J|$ at 25 kHz within muscle tissue (total volume: 23.2 l)

A.1.3 Simulations $f = 50$ kHz, $|\mathbf{B}|_{\mathbf{center}}$ = 10 mT

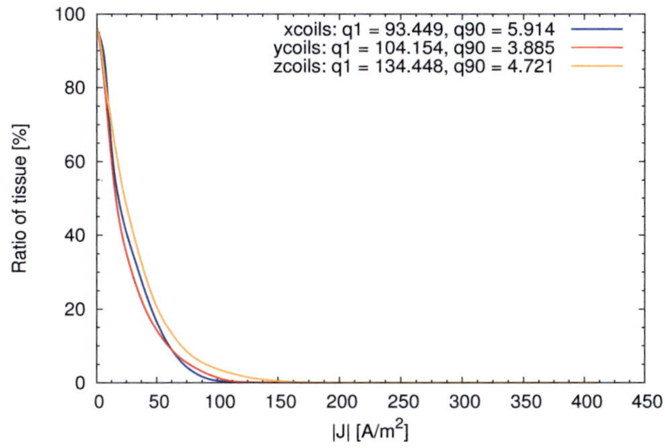

Fig. A.7. Cumulative histogram of induced $|J|$ at 50 kHz within the entire torso (total volume: 65 l)

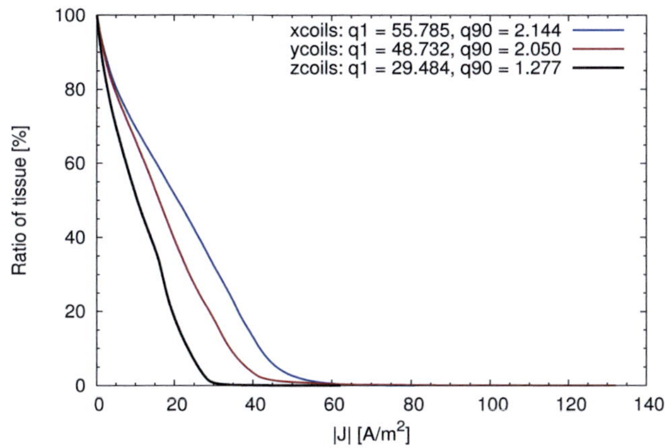

Fig. A.8. Cumulative histogram of induced $|J|$ at 50 kHz within the heart (total volume: 0.35 l)

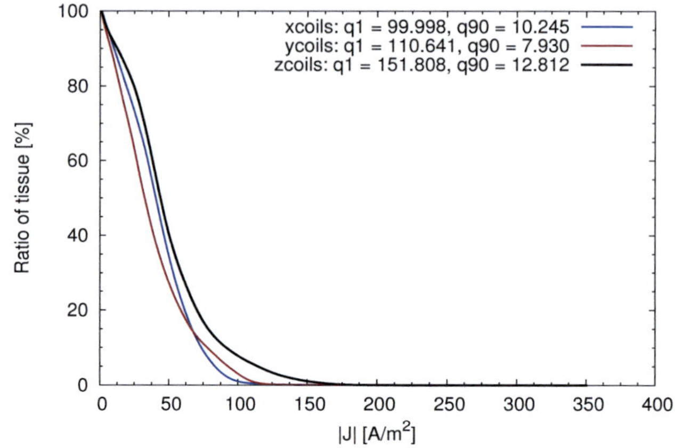

Fig. A.9. Cumulative histogram of induced $|J|$ at 50 kHz within muscle tissue (total volume: 23.2l)

A.1.4 Simulations $f = 100$ kHz, $|\mathbf{B}|_{\mathbf{center}}$ = 10 mT

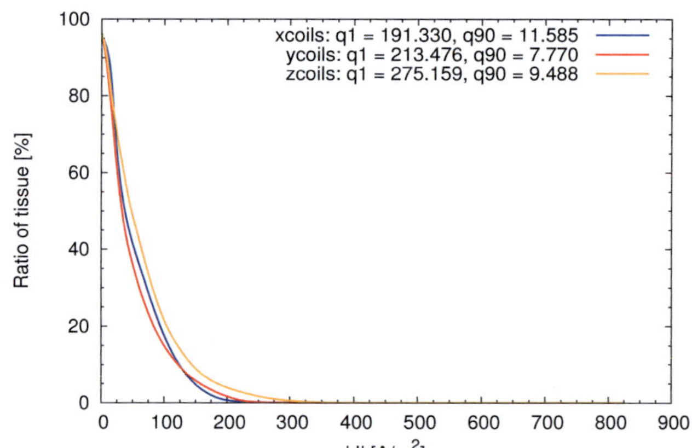

Fig. A.10. Cumulative histogram of induced $|J|$ at 100 kHz within the entire torso (total volume: 65 l)

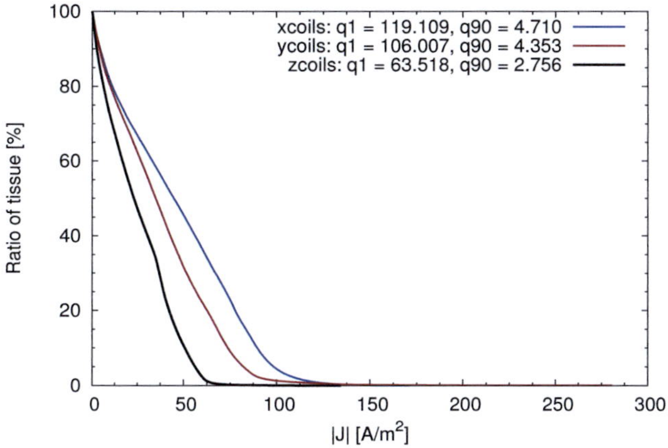

Fig. A.11. Cumulative histogram of induced $|J|$ at 100 kHz within the heart (total volume: 0.35 l)

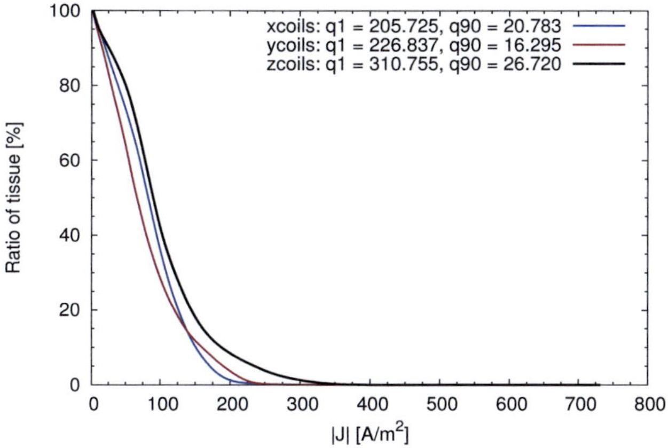

Fig. A.12. Cumulative histogram of induced $|J|$ at 100 kHz within muscle tissue (total volume: 23.2l)

A.1.5 Magnetic Stimulation: Current Density induced in the Human Arm Cumulative Histogram

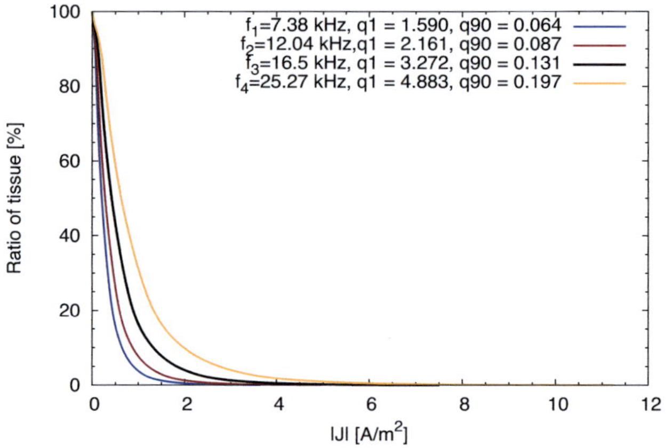

Fig. A.13. Cumulative histogram of induced $|J|$ at four stimulation frequencies within muscle tissue (total volume: 1.5 l)

List of Figures

List of Tables

References

1. M. Brand and O. Heid, "Induction of electric fields due to gradient switching: a numerical approach," *Magn Reson Med*, vol. 48, no. 4, pp. 731–734, 2002.
2. T. F. Budinger, H. Fischer, D. Hentschel, H.-E. Reinfelder, and F. Schmitt, "Physiological effects of fast oscillating magnetic field gradients," *J Comp Assist Tomogr*, vol. 15, pp. 909–914, 1991.
3. L. Forbes and S. Crozier, "On a possible mechanism for peripheral nerve stimulation during magnetic resonance imaging scans," *Phys Med Biol*, vol. 46, p. 591, 2001.
4. P. So, M. Stuchly, and J. Nyenhuis, "Peripheral nerve stimulation by gradient switching fields in magnetic resonance imaging," *IEEE Trans Biomed Eng*, vol. 51, no. 11, pp. 1907–1914, 2004.
5. S. Seitz, *Magnetic resonance imaging on patients with implanted cardiac pacemakers*. PhD thesis, Institute of Biomedical Engineering, Karlsruhe Institute of Technology, 2011.
6. P. J. Dimbylow, "Induced current densities from low-frequency magnetic fields in a 2 mm resolution, anatomically realistic model of the body," *Phys Med Biol*, vol. 43, pp. 221–230, 1998.
7. J. A. Ambrose and G. Hounsfield, "Computerized transverse axial tomography," *Brit J Radiol*, vol. 46, pp. 148–149, 1973.
8. G. Hounsfield, "Computerized transverse axial scanning (tomography) - Part I. Description of system. 1973," *Brit J Radiol*, vol. 68, pp. H116–H172, 1995.
9. P. C. Lauterbur, "Image formation by induced local interactions: examples employing nuclear magnetic resonance," *Nature*, vol. 242, pp. 190–191, 1973.
10. M. M. Ter-Pogossian, M. E. Raichle, and B. E. Sobel, "Positron-emission tomography," *Scientific America*, vol. 243, pp. 170–181, 1980.
11. B. Gleich and J. Weizenecker, "Tomographic imaging using the nonlinear response of magnetic particles," *Nature*, vol. 435, pp. 1214–1217, 2005.
12. B. Gleich, J. Weizenecker, and J. Borgert, "Experimental results on fast 2D-encoded magnetic particle imaging," *Phys Med Biol*, vol. 53, p. 81, 2008.
13. J. Weizenecker, J. Borgert, and B. Gleich, "A simulation study on the resolution and sensitivity of magnetic particle imaging," *Phys Med Biol*, vol. 52, pp. 6363–6374, 2007.
14. J. Weizenecker, B. Gleich, J. Rahmer, H. Dahnke, and J. Borgert, "Three-dimensional real-time in vivo magnetic particle imaging," *Phys Med Biol*, vol. 54, pp. L1–L10, 2009.
15. T. Knopp, J. Borgert, and T. M. Buzug, eds., *Magnetic Nanoparticles: Particle Science, Imaging Technology, and Clinical Applications*, vol. 1. World Scientific Publishing Company, 2010.
16. T. Knopp, S. Biederer, T. Sattel, J. Weizenecker, B. Gleich, J. Borgert, and T. M. Buzug, "Trajectory analysis for magnetic particle imaging," *Phys Med Biol*, vol. 54, no. 2, p. 385, 2009.

17. T. Knopp, *Effiziente Rekonstruktion und alternative Spulentopologien für Magnetic-Particle-Imaging*. PhD thesis, Institut für Medizintechnik, Universität zu Lübeck, 2011.
18. A. Dax, "On Row Relaxation Methods for Large Constrained Least Squares Problems," *SIAM J Sci Comput*, vol. 14, no. 3, pp. 570–584, 1993.
19. C. Furse, C. H. Durney, and D. A. Christensen, *Basic Introduction to Bioelectromagnetics*. Crc Pr Inc, 2009.
20. M. Sadiku, *Elements of electromagnetics*. Oxford university press, 2001.
21. A. J. Schwab and F. Imo, *Begriffswelt der Feldtheorie*. Berlin, Heidelberg: Springer, 2002.
22. SPEAG (Schmid & Partner Engineering AG), *SEMCAD X Reference Manual (Ver. 14.2)*. Zürich, Switzerland, 2010.
23. S. Silbernagl and A. Despopoulos, *Taschenatlas der Physiologie*, ch. Grundlagen, Zellphysiologie, pp. 2–41. Stuttgart: Georg Thieme Verlag, 2001.
24. J. Dudel, *Physiologie des Menschen*, ch. Grundlagen der Zellphysiologie, pp. 3–19. Berlin, Heidelberg, New York: Springer, 2000.
25. R. F. Schmidt, G. Thews, and F. Lang, *Physiologie des Menschen*. Berlin; Heidelberg; New York: Springer, 2000.
26. H. P. Schwan, "Electrical properties of tissue and cell suspensions," *Adv Biol Med Phys*, vol. 5, pp. 147–209, 1957.
27. K. S. Cole and R. H. Cole, "Dispersion and absorption in dielectrics. I. Alternating current characteristics," *J Chem Phys*, vol. 9, pp. 341–351, 1941.
28. S. Gabriel, R. W. Lau, and C. Gabriel, "The dielectric properties of biological tissues: III. Parametric models for the dielectric spectrum of tissues," *Phys Med Biol*, vol. 41, pp. 2271–2293, 1996.
29. M. A. Golombeck, *Feldtheoretische Studien zur Patientensicherheit bei der Magnetresonanztomographie und der Elektrochirugie*. PhD thesis, Universität Karlsruhe, (TH), 2003.
30. G. W. Beeler and H. Reuter, "Voltage clamp experiments on ventricular myocardial fibres," *J Physiol*, pp. 165–190, 1970.
31. A. L. Hodgkin and A. F. Huxley, "A quantitative description of membrane current and its application to conduction and excitation in nerve," *J Physiol*, vol. 117, pp. 500–544, 1952.
32. C.-H. Luo and Y. Rudy, "A model of the ventricular cardiac action potential. Depolarization, repolarization, and their interaction," *Circ res*, vol. 68, pp. 1501–1526, 1991.
33. C. H. Luo and Y. Rudy, "A dynamic model of the cardiac ventricular action potential. I. Simulations of ionic currents and concentration changes," *Circ Res*, vol. 74, pp. 1071–1096, 1994.
34. L. Priebe and D. J. Beuckelmann, "Simulation Study of Cellular Electric Properties in Heart Failure," *Circ Res*, vol. 82, pp. 1206–1223, 1998.
35. K. H. W. J. ten Tusscher, D. Noble, P. J. Noble, and A. V. Panfilov, "A model for human ventricular tissue," *Am J Physiol Heart C*, vol. 286, pp. H1573–89, 2004.
36. K. H. W. J. Ten Tusscher and A. V. Panfilov, "Cell model for efficient simulation of wave propagation in human ventricular tissue under normal and pathological conditions," *Phys Med Biol*, vol. 51, pp. 6141–6156, 2006.
37. C. A. Otto, *Impact of adrenergic regulation on the electrophysiological properties of human ventricular myocytes*. diploma thesis, Institute of Biomedical Engineering, Karlsruhe Institute of Technology (KIT), 2009.
38. J. P. Reilly, "Peripheral nerve stimulation by induced electric currents: exposure to time-varying magnetic fields," *Med Biol Eng Comp*, vol. 27, pp. 101–110, 1989.

39. R. J. Ilmoniemi, J. Ruohonen, and J. Karhu, "Transcranial magnetic stimulation–a new tool for functional imaging of the brain," *Crit Rev Biomed Eng*, vol. 27, pp. 241–284, 1999.
40. B. J. Roth, "Mechanisms for electrical stimulation of excitable tissue," *Crit Rev Biomed Eng*, vol. 3, pp. 253–305, 1994.
41. B. J. Roth and P. J. Basser, "A model of the stimulation of a nerve fiber by electromagnetic induction," *IEEE Trans biomed Eng*, vol. 37, pp. 588–597, 1990.
42. H. A. Blair, "On the intensity-time relations for stimulation by electric currents. I," *J Gen Physiol*, vol. 15, pp. 709–729, 1932.
43. J. P. Reilly, *Electrical stimulation and electropathology*. Cambridge: Cambridge University Press, 1992.
44. J. Ranck *et al.*, "Which elements are excited in electrical stimulation of mammalian central nervous system: a review," *Brain Research*, vol. 98, no. 3, pp. 417–440, 1975.
45. J. P. Reilly and R. H. Bauer, "Application of a neuroelectric model to electrocutaneous sensory sensitivity: parameter variation study," *IEEE Trans BioMed Eng*, vol. 34, pp. 752–754, 1987.
46. J. P. Reilly, "Neuroelectric mechanisms applied to low frequency electric and magnetic field exposure guidelines–part I: sinusoidal waveforms," *Health Phys*, vol. 83, pp. 341–355, 2002.
47. J. P. Reilly, "Magnetic field excitation of peripheral nerves and the heart: a comparison of thresholds," *Med Biol Eng Comp*, vol. 29, pp. 571–579, 1991.
48. W. Irnich, "Electrostimulation by time-varying magnetic fields," *Magn Reson Mater Phy*, vol. 2, pp. 43–49, 1994.
49. O. A. Wyss, "Mittelfrequenzreizung des Nerven bei Querdurchströmung," *Pflügers Archiv*, vol. 274, p. 94, 1961.
50. O. A. Wyss, "Das apolaritäre Prinzip der Mittelfrequenz-Reizung," *Experientia*, vol. 23, pp. 601–696, 1967.
51. B. Bromm and H. Lullies, "Über den Mechanismus der Reizwirkung mittelfrequenter Wechselströme auf die Nervenmembran," *Pflügers Archiv Europ J Physiol*, vol. 289, pp. 215–226, 1966.
52. I. Chatterjee, D. Wu, and O. P. Gandhi, "Human body impedance and threshold currents for perception and pain for contact hazard analysis in the VLF-MF band," *IEEE Trans biomed Eng*, vol. 33, pp. 486–494, 1986.
53. L. A. Geddes and L. E. Baker, "Response to passage of electric current through the body," *J Assoc Adv Med Instrum*, vol. 5, pp. 13–18, 1971.
54. D. McRobbie and M. A. Foster, "Thresholds for biological effects of time-varying magnetic fields," *Clin Phys Physiol Meas*, vol. 5, pp. 67–78, 1984.
55. M. Polson, A. Barker, and S. Gardiner, "The effect of rapid rise-time magnetic fields on the ECG of the rat," *Clin Phys Physiol Meas*, vol. 3, p. 231, 1982.
56. M. Polson, A. Barker, and I. Freeston, "Stimulation of nerve trunks with time-varying magnetic fields," *Med Biol Eng Comput*, vol. 20, pp. 243–244, 1982.
57. S. Ueno, "Biological effects of magnetic fields," *IEEE Transl J Magn Japan*, vol. 7, pp. 580–585, 1992.
58. D. Irwin, S. Rush, R. Evering, E. Lepeschkin, D. Montgomery, and R. Weggel, "Stimulation of cardiac muscle by a time-varying magnetic field," *IEEE Trans Magn*, vol. 6, pp. 321–322, 1970.
59. H. H. Pennes, "Analysis of tissue and arterial blood temperatures in the resting human forearm," *J Appl Physiol*, vol. 1, pp. 93–122, 1948.
60. ICNIRP (International Commission on Non-Ionizing Radiation Protection, "Guidelines for limiting exposure to time-varying electric, magnetic, and electromagnetic fields (up to 300 GHz)," *Health Phys*,

vol. 74, pp. 494–522, 1998.

61. ICNIRP (International Commission on Non-Ionizing Radiation Protection, "Guidelines for limiting exposure to time-varying electric and magnetic fields (1 Hz To 100 kHz)," *Health Phys*, pp. 818–836, 2010.
62. INIRC (International Non-Ionizing Radiation Committee), "Protection of the patient undergoing a magnetic resonance examination," *Health Phys*, vol. 61, pp. 923–928, 1991.
63. National Library of Medicine, Bethesda, USA, "Visible Human Project."
64. M. J. Ackerman, V. M. Spitzer, A. L. Scherzinger, and D. G. Whitlock, "The Visible Human data set: an image resource for anatomical visualization," *Med info*, vol. 8, pp. 1195–1198, 1995.
65. V. Spitzer, M. J. Ackerman, A. L. Scherzinger, and D. Whitlock, "The Visible Human Male: A Technical Report," *J Am Med Inform Assoc*, vol. 3, pp. 118–130, 1996.
66. C. Gabriel, S. Gabriel, and E. Corthout, "The dielectric properties of biological tissues: I. Literature survey," *Phys Med Biol*, vol. 41, pp. 2231–2250, 1996.
67. S. Gabriel, R. W. Lau, and C. Gabriel, "The dielectric properties of biological tissues: II. Measurements in the frequency range 10 Hz to 20 GHz," *Phys Med Biol*, vol. 41, pp. 2251–2269, 1996.
68. D. U. J. Keller, F. M. Weber, G. Seemann, and O. Dössel, "Ranking the influence of tissue conductivities on forward-calculated ECGs," *IEEE Trans Biomed Eng*, vol. 57, pp. 1568–1576, 2010.
69. IEEE, "Recommended practice for measurements and computations of radio frequency electromagnetic fields with respect to human exposure to such fields,100 kHz–300 GHz," *IEEE Std C95.3-2002 (Revision of IEEE Std C95.3-1991)*, p. 126, 2002.
70. B. J. Roth, L. G. Cohen, M. Hallett, W. Friauf, and P. J. Basser, "A theoretical calculation of the electric field induced by magnetic stimulation of a peripheral nerve," *Muscle & Nerve*, vol. 13, pp. 734–741, 1990.
71. J. P. Reilly, *Electrical stimulation and electropathology*. Cambridge: Cambridge University Press, 1. publ. ed., 1992.
72. C. M. Collins, W. Liu, J. Wang, R. Gruetter, J. T. Vaughan, K. Ugurbil, and M. B. Smith, "Temperature and SAR calculations for a human head within volume and surface coils at 64 and 300 MHz," *J Magn Reson Imag*, vol. 19, pp. 650–656, 2004.
73. M. A. Golombeck, C. H. Riedel, and O. Dössel, "Calculation of the Dielectric Properties of Biological Tissue using simple Models of Cell Patches," in *Biomed Tech*, vol. 47-1, pp. 253–256, 2002.
74. A. Stogryn, "Equations for calculating the dielectric constant of saline water," *IEEE Trans Microwav Theory*, vol. 19, pp. 733–736, 1971.
75. O. Bernus, R. Wilders, C. W. Zemlin, H. Verschelde, and A. V. Panfilov, "A computationally efficient electrophysiological model of human ventricular cells," *Am J Physiol*, vol. 282, p. 2296, 2002.
76. E. J. Speckmann, H. J., and K. R., *Physiologie*. Elsevier GmbH, Urban & Fischer Verlag, 2008.
77. Kuypers Kliniek, *www.handenpolscentrumnoordholland.nl/compressie-nervus-ulnaris.php*. 1625 HV Hoorn, The Netherlands, 2009.
78. L. G. Cohen, B. J. Roth, J. Nilsson, N. Dang, M. Panizza, S. Bandinelli, W. Friauf, and M. Hallett, "Effects of coil design on delivery of focal magnetic stimulation. Technical considerations," *Electroen Clin Neuro*, vol. 75, pp. 350–357, 1990.
79. K. H. Hsu and D. M. Durand, "A 3-D differential coil design for localized magnetic stimulation," *IEEE Trans biomed Eng*, vol. 48, pp. 1162–1168, 2001.
80. H. Wheeler, "Simple Inductance Formulas for Radio Coils," *Proc Inst Radio Eng*, vol. 16, pp. 1398 – 1400, oct. 1928.

81. M. Gräser, *Bestimmung von Schwellwerten zur induktiven Muskelstimulation im Bereich von 7,4 kHz bis 25 kHz*. diploma thesis, Institute of Biomedical Engineering, Karlsruhe Institute of Technology (KIT), 2010.

Karlsruhe Transactions on Biomedical Engineering (ISSN 1864-5933)

Karlsruhe Institute of Technology / Institute of Biomedical Engineering (Ed.)

Die Bände sind unter www.ksp.kit.edu als PDF frei verfügbar oder als Druckausgabe bestellbar.

Band 2 Matthias Reumann
Computer assisted optimisation on non-pharmacological treatment of congestive heart failure and supraventricular arrhythmia. 2007
ISBN 978-3-86644-122-4

Band 3 Antoun Khawaja
Automatic ECG analysis using principal component analysis and wavelet transformation. 2007
ISBN 978-3-86644-132-3

Band 4 Dmytro Farina
Forward and inverse problems of electrocardiography : clinical investigations. 2008
ISBN 978-3-86644-219-1

Band 5 Jörn Thiele
Optische und mechanische Messungen von elektrophysiologischen Vorgängen im Myokardgewebe. 2008
ISBN 978-3-86644-240-5

Band 6 Raz Miri
Computer assisted optimization of cardiac resynchronization therapy. 2009
ISBN 978-3-86644-360-0

Band 7 Frank Kreuder
2D-3D-Registrierung mit Parameterentkopplung für die Patientenlagerung in der Strahlentherapie. 2009
ISBN 978-3-86644-376-1

Band 8 Daniel Unholtz
Optische Oberflächensignalmessung mit Mikrolinsen-Detektoren für die Kleintierbildgebung. 2009
ISBN 978-3-86644-423-2

Band 9 Yuan Jiang
Solving the inverse problem of electrocardiography in a realistic environment. 2010
ISBN 978-3-86644-486-7

Karlsruhe Transactions on Biomedical Engineering (ISSN 1864-5933)

Band 10 Sebastian Seitz
Magnetic Resonance Imaging on Patients with Implanted Cardiac Pacemakers. 2011
ISBN 978-3-86644-610-6

Band 11 Tobias Voigt
Quantitative MR Imaging of the Electric Properties and Local SAR based on Improved RF Transmit Field Mapping. 2011
ISBN 978-3-86644-598-7

Band 12 Frank Michael Weber
Personalizing Simulations of the Human Atria: Intracardiac Measurements, Tissue Conductivities, and Cellular Electrophysiology. 2011
ISBN 978-3-86644-646-5

Band 13 David Urs Josef Keller
Multiscale Modeling of the Ventricles: from Cellular Electrophysiology to Body Surface Electrocardiograms. 2011
ISBN 978-3-86644-714-1

Band 14 Oussama Jarrousse
Modified Mass-Spring System for Physically Based Deformation Modeling. 2012
ISBN 978-3-86644-742-4

Band 15 Julia Bohnert
Effects of Time-Varying Magnetic Fields in the Frequency Range 1 kHz to 100 kHz upon the Human Body: Numerical Studies and Stimulation Experiment. 2012
ISBN 978-3-86644-782-0